Organic Semiconductors and Biopolymers

MONOGRAPHS IN SEMICONDUCTOR PHYSICS

1969

Volume 1
HEAVILY DOPED SEMICONDUCTORS
by Viktor I. Fistul'

Volume 2
LIQUID SEMICONDUCTORS
by V. M. Glazov, S. N. Chizhevskaya, and N. N. Glagoleva

Volume 3
SEMICONDUCTING II-VI, IV-VI, and V-VI COMPOUNDS
by N. Kh. Abrikosov, V. F. Bankina, L. V. Poretskaya, L. E. Shelimova, and E. V. Skudnova

Volume 4
SWITCHING IN SEMICONDUCTOR DIODES
by Yu. R. Nosov

1970

Volume 5
SEMICONDUCTING LEAD CHALCOGENIDES
by Yu. I. Ravich, B. A. Efimova, and I. A. Smirnov

Volume 6
ORGANIC SEMICONDUCTORS AND BIOPOLYMERS
by L. I. Boguslavskii and A. V. Vannikov

Organic Semiconductors and Biopolymers

Leonid I. Boguslavskii
and
Anatolii V. Vannikov
Institute of Electrochemistry
Academy of Sciences of the USSR
Moscow, USSR

Translated from Russian by
B. J. Hazzard

PLENUM PRESS • NEW YORK–LONDON • 1970

Leonid Isaakovich Boguslavskii was born in 1935. In 1958 he graduated from the chemical faculty of Moscow University. Since 1958 he has been working in the Institute of Electrochemistry of the Academy of Sciences of the USSR. His Candidate's thesis, defended in 1964, was devoted to the mechanism of conduction and the surface properties of complexes of tetracyanoethylene with metals. He has carried out a number of investigations of the surface properties of organic semiconductors. In particular, he has studied the influence of the work function of an electron from the surface of an organic semiconductor on the adsorbability of acceptor gases. The fields which interest Boguslavskii at present are adsorption from solutions onto the surface of single crystals of anthracene and the study of the proton conductivity of synthetic phospholipid membranes.

Anatolii Veniaminovich Vannikov was born in 1937 and graduated from the chemical faculty of Moscow University in 1959. Since then he has worked in the Institute of Electrochemistry of the Academy of Sciences of the USSR. He defended his Candidate's thesis in 1965. He has worked in the field of polymeric organic semiconductors, his main interests being the study of the electrophysical properties of polymeric semiconductors and the mechanism of conduction in them.

The original Russian text, published by Nauka Press in 1968 for the Institute of Electrochemistry of the Academy of Sciences of the USSR, has been corrected by the authors for the present edition. The English translation is published under an agreement with Mezhdunarodnaya Kniga, the Soviet book export agency.

Леонид Исаакович Богуславский,
Анатолий Вениаминович Ванников

Органические полупроводники и биополимеры

ORGANICHESKIE POLUPROVODNIKI I BIOPOLIMERY

ORGANIC SEMICONDUCTORS AND BIOPOLYMERS

Library of Congress Catalog Card Number 72-75452

SBN 306-30433-3

A Division of Plenum Publishing Corporation
227 West 17th Street, New York, N.Y. 10011

United Kingdom edition published by Plenum Press, London
A Division of Plenum Publishing Corporation, Ltd.
Donington House, 30 Norfolk Street, London W.C.2, England

Printed in the United States of America

PREFACE TO THE AMERICAN EDITION

If we compare Soviet and foreign scientific literature on organic semiconductors, it is impossible not to observe that this region of knowledge is a peculiar example of the international division of labor. While the majority of Soviet investigations has been devoted to the study of polymeric organic semiconductors, foreign authors — for example, American ones — have worked with molecular crystals. And although it is possible to find first-class individual investigations in the opposite field on both sides, we nevertheless hope that an acquaintance with Soviet work in the field of polymeric organic semiconductors will be of interest for those working in this field of science.

The Authors

FOREWORD

In recent years, considerable progress has been made in the study of organic semiconductors. The main directions of investigation have been determined and substantial results have been achieved in the theoretical treatment of many questions. However, the range of investigations is so broad that it is impossible to discuss all branches fully in a short monograph. In fact, the chemist synthesizing systems with conjugated bonds and studying their reactivity, the physical chemist studying the catalytic and absorption properties of substances with conjugated bonds, the physicist studying the photoelectric and magnetic characteristics of dyes and polynuclear hydrocarbons, and the biologist attempting to explain the mechanism of energy transfer in the living cell — they are all working in the field of organic semiconductors. Consequently, in the present book the authors have not attempted to include all sides of the problem of organic semiconductors. The monograph considers in detail only those questions which appear to the authors to be most immediate and interesting. Other questions are touched on only to the extent necessary to illustrate the main material.

The book begins with a statement of material which is of the nature of a review. The main classes of organic semiconductors are mentioned but the methods for their preparation are not illustrated in detail. Somewhat more attention is devoted to the dependence of the electrical characteristics of organic semiconductors on their structure and chemical composition.

Photoelectric characteristics are considered only for polymeric organic semiconductors, since there is a voluminous literature on the photoconductivity of dyes and polynuclear hydrocarbons. In the discussion of the magnetic properties of systems with conjugated bonds, our main attention is devoted to comparing their electrical and paramagnetic characteristics.

A large part of the book is devoted to considering the mechanism of conduction in organic semiconductors. While in low-molecular-weight crystals the main features of the mechanism of conduction are known, in polymeric semiconductors, because of the complexity of their structure, the mechanism of conduction is far from clear. This is connected, in particular, with the fact that the methods of measuring the mobility of charge carriers that are traditional in semiconductor physics are not suitable for polymeric semiconductors. Consequently, the book gives a detailed consideration of methods permitting the direct measurement of the mobility of the carriers.

In their account of the surface properties of organic semiconductors, the authors stress the great role of the state of the surface and of the absorption of various gases on the observed electrical and magnetic properties. So far as the catalytic activity of organic semiconductors is concerned, the main problem has been to compare the observed catalytic properties with other parameters.

In considering the electrical properties of biopolymers, ideas developed to explain the mechanism of conduction in polymeric organic semiconductors have been used. The authors hope that this approach will prove to be extremely promising for an explanation of many elementary processes in biological systems.

In the concluding part of the book, the authors attempt to give as fully as possible data relating to the properties of the practical utilization of organic semiconductors.

The authors take this opportunity to express their deep gratitude to A. M. Kuznetsov, E. V. Sidorova, Yu. A. Chizmadzhev, and R. M. Vlasova for useful discussions and valuable observations.

The Authors

CONTENTS

Chapter I

Chapter II

Chapter III

Chapter IV

Chapter V

Chapter VI

Chapter VII

Chapter VIII

CHAPTER 1

PREPARATION, CONSTRUCTIVE PROPERTIES, AND MAIN ELECTRICAL CHARACTERISTICS OF ORGANIC SEMICONDUCTORS

One of the most important problems of present-day chemistry is the creation of new substances and materials possessing a series of valuable properties. Particularly great prospects have been opened up in the synthesis and study of organic compounds possessing delocalized electrons because of the presence in them of conjugated bonds or the formation of charge-transfer complexes. Such compounds have acquired the name of organic semiconductors. Starting in 1960, the volume of investigations on the chemistry and physics of these materials has risen to such an extent that the possibility has appeared of generalizing some basic achievements in this field [1, 2]. At the present time organic semiconductors include both low-molecular-weight compounds and polymers with a large number of conjugated bonds, charge-transfer complexes in which, owing to a definite structure, conditions are created for the delocalization of an electron, stable free radicals, and also some biopolymers (including proteins) the conductivity of which can be explained by a nonionic mechanism.

The greatest number of organic semiconductors, both low-molecular-weight compounds and polymers, has been obtained by synthesis. Of the low-molecular-weight compounds, the best-known are aromatic (polycyclic and nitrogen-containing) compounds, dyes, and phthalocyanines.

According to the classification proposed by Berlin [3], polymeric materials with a system of conjugation can be divided into three groups.

POLYMERS WITH LINEAR CONJUGATION IN THE MAIN CHAIN

To the first group must be assigned the polyvinylenes, i.e., polymers with an acyclic system of conjugation. In such molecules, free rotation about the C=C bond is limited and this determines the planar arrangement of the whole chain. When the polymer chain is of appreciable length, there may be a wide choice of conformational structures.

Such structures possess a greater density and higher boiling and melting points than the corresponding hydrocarbons containing isolated double bonds. The delocalization of the π-electrons along the chain of conjugation is responsible for their high heat-stability. The increased chemical stability of such compounds is shown by the fact that they are fairly inert to addition reactions. As an example we may give the most important characteristics of polymers based on phenylacetylene. The mean molecular weight of the polyphenylacetylenes is 1100-1500. The absorption spectra of the polyphenylacetylenes show that they consist of fractions with different molecular weights. The longest macromolecules have an absorption spectrum in the visible region. The polyphenylacetylenes give a narrow EPR signal and, depending on their molecular weight, have colors from yellow to black and contain 10^{17}-10^{20} paramagnetic particles per gram.

POLYMERS WITH AROMATIC NUCLEI IN THE CHAIN OF CONJUGATION

This type of polymer includes organic semiconductors in which the continuity of the chain of conjugation is ensured by aromatic nuclei or groups included directly in the main chain. Polyphenylene [3] is an example of this type of polymer.

Depending on their method of preparation, the molecular weights of such polymers may differ. Soluble branched polymers can be obtained with molecular weights of 3000-4000 and may contain 10^{17}-10^{19} paramagnetic particles per gram. Such polymers are stable at a temperature of 500-600°C and are dielectrics at room temperature. However, on heating, their conductivity rises sharply, since the activation energy of electrical conductivity is 1.5 eV [4]. Polyphenylenes with high molecular weights are infusible and insoluble and their electrical conductivity is 3-4 orders of magnitude greater than that of the low-molecular-weight compounds.

Polyazophenylenes, polyphenylenequinones, and polyphenyleneferrocenes with various functional groups may be assigned to the same type of polymers [3, 5, 6]. The polyazophenylenes are extremely heat-stable. In the absence of oxygen, the loss in weight at 500-700°C is 4-6%. In air, these polymers lose 20% of their weight at 450°C. Polyazophenylenes containing carboxy groups and sulfo groups possess cation-exchange properties [4].

POLYMERS WITH HETEROCYCLES AND METALLOCYCLES IN THE CHAIN OF CONJUGATION

Of these polymers, the best-studied is thermally modified polyacrylonitrile (PAN) consisting of a system of condensed pyridine rings [8], polymeric phthalocyanines [9], complexes of tetracyanoethylene with metals [10], polymeric Schiff bases, and polyazines [7, 11].

Thermally treated polyacrylonitrile is formed by heating the initial polyacrylonitrile, which is converted into a dark-colored material at about 200-300°C. Depending on its thermal treatment, its electrical properties vary over an extremely wide range. With an increase in the temperature of the treatment, the specific resistance falls from that characteristic for an insulator down to 1-2 Ω · cm, and the activation energy of electrical conductivity from one and a half to several hundred eV. Thermally treated polyacrylonitrile is prepared in the form of films, fibers, and fabrics. Polymeric phthalocyanines and complexes of tetracyanoethylene with metals are obtained in powdered form, but they can be prepared as films through the reaction of the vapors of the starting material with a metallic surface. The metal atoms enter into the composition of the film, which proves to be chemically bound to the substrate. The specific resistance of such films varies from 10^7 to $10^2\ \Omega\cdot$cm, depending on the metal and the method of preparation, and the activation energy of electrical conductivity amounts to several tens of eV. Very frequently the curve of the resistance as a function of the reciprocal temperature shows a break corresponding to two activation energies of electrical conduction, as is the case in impurity (extrinsic) semiconductors.

POLYMERS WITH A SYSTEM OF CONJUGATED DOUBLE BONDS OBTAINED BY THERMAL AND RADIATION METHODS

One of the methods of obtaining polyconjugated systems is the creation of changes in the chains of saturated macromolecules under the action of thermal, radiation, and other factors.

This method was first used by Winslow et al. [12, 13], who studied the dependence of the electrophysical properties of thermally treated polydivinylbenzene and poly(vinyl chloride).

The splitting out of atoms or molecules from saturated polymeric structures at a high temperature leads to the formation of materials with polyconjugated systems possessing various electrophysical properties. The most likely process taking place under these conditions is, as has been shown by Grassie [14],

$$\left[\begin{array}{c} \mathrm{H}\quad\mathrm{X} \\ |\quad\ | \\ -\mathrm{C}-\mathrm{C}- \\ |\quad\ | \\ \mathrm{H}\quad\mathrm{Y} \end{array}\right]_n \longrightarrow \left[\begin{array}{c} -\mathrm{C}=\mathrm{C}- \\ |\quad\ | \\ \mathrm{H}\quad\mathrm{Y} \end{array}\right]_n + n\,\mathrm{HX},$$

where X=Cl, Br, OH, $\mathrm{CH_3{-}\underset{\underset{\displaystyle O}{\|}}{C}{-}O}$, etc.

By this method, polymers with a system of conjugation have been obtained from poly(vinyl chloride) [15-17], poly(vinylidene chloride) [12, 13, 18], poly(vinyl bromide) [19, 20], poly(vinyl acetate) [21], and poly(vinyl alcohol) [22, 23]. The possibility of obtaining semiconducting materials by the thermal treatment of ion-exchange resins is discussed by Pohl [24, 25].

The best-studied of these processes is the conversion of polyacrylonitrile into a polymer with conjugated double bonds on heat treatment [1, 26, 27]. The thermal treatment of polyacrylonitrile leads to the formation of six-membered rings with the splitting out of hydrogen molecules and the appearance of a polyconjugated system. Depending on the temperature of treatment, the electrophysical properties of the materials obtained vary within very wide limits. The structural transformations taking place in the thermal treatment of polyacrylonitrile have been studied by Kargin and Litvinov [28].

It has been shown by electron-microscope and x-ray radiographic methods that at temperatures up to 400°C there are no visible changes whatever in the structure and the boundaries of separation of the spherulites and the fibrillar nature of the supermolecular structure is well preserved. The supermolecular structure is preserved in treatment even at temperatures up to 800°C. This shows that the structural changes which polyacrylonitrile undergoes during thermal treatment leads to a polymeric material having a cyclic structure with the retention of the supermolecular structure of the initial crystalline sample, i.e., there is the occurrence of pseudomorphism — the retention of the external form but a change in the internal structure consisting in the accumulation of conjugated double bonds.

Similar investigations clearly show that previously oriented samples retain the predominant direction of the macromolecules even after thermal treatment. This opens up a real route to obtaining semiconducting polymeric materials with anisotropic conductivity.

In addition to the thermal treatment of the polymer, the influence of intensive infrared radiation on polyacrylonitrile has been studied [29, 30]. In this case, a selective action of the radiation on the vibrational energy of the individual groups of macromolecules was observed. Structural transformations of the polyacrylonitrile took place with the formation of conjugated $(-N=C-)_n$ bonds throughout the macromolecular chain.

Under the action of ultraviolet radiation on copolymers of vinyl chloride and vinylidene chloride [31], an increase in conductivity accompanied by the coloration of the polymer was found. Higher energy treatment of the polymer also leads to the creation of a system of conjugation. It was found [32] that at equivalent absorbed doses γ-radiation and the mixed γ,n-radiation of a nuclear reactor were completely identical in their effect on the electrophysical properties of polymers.

The combination of radiation and thermal effects also leads to the formation of polymers with semiconducting properties [33, 34]. This method has proved successful in the production of highly conducting films from such readily available linear polymers as polyethylene and poly(vinyl acetate).

The existence of a system of conjugated bonds in a polymeric molecule leads to an increase in its thermodynamic stability which is shown, in particular, in the thermal stability of the polymers.

THERMAL STABILITY OF POLYMERS

A polymeric material used at high temperatures (>300°C) must satisfy certain demands that have been formulated in the literature [35, 36]: high melting or softening point, high resistance to spontaneous pyrolysis; considerable resistance to degradation under the influence of chemical reagents.

When high temperatures act on a polymer, two processes that change the chemical properties of the polymer through chemical transformations may take place. The first – the degradation of the polymer – leads to a gradual decrease in the molecular weight resulting from the cleavage of the chains of the molecules and is accompanied by a loss in strength and rigidity and a decrease in the softening temperature. The second process – cross-linkage – increases the molecular weight because of the formation of bonds between separate chains, so that a rigid cross-linked structure which is infusible and insoluble is created.

Generally, both processes take place simultaneously. The higher the limiting temperature at which both processes begin, the greater is the thermal stability of the polymer. Cleavage and the formation of new bonds begins at a temperature where the thermal energy becomes comparable with the energy of cleavage of the bonds in the polymers. The kinetics of the chemical changes in the polymer are generally studied from the rate of evolution of gases, infrared spectra, and the loss in weight when the polymer is heated. The kinetics of the degradation of polymers with conjugated bonds have a completely different character from the kinetics of compounds with ordinary bonds or isolated double bonds. For conjugated system, the polymer undergoes a rapid loss in weight during the initial period of degradation and then the change takes place more slowly. If the stable residue obtained is heated to a higher temperature, a relatively rapid change in weight and subsequent stabilization again take place.

This process may be repeated several times, each rise in the temperature corresponding to its step on the degradation curve (Fig. 1) [37, 38], which shows that when the temperature is raised the molecules undergo some changes or other which lead to their stabilization. The result of the process is the formation of carbonized or graphitized structures at 2000-3000°C. The explanation of this self-stabilization is obviously that as the temperature is raised an increase in the total system of conjugation takes place and, consequently, there is a decrease in the internal energy of the macromolecules. The formation of structures with increasingly favorable energies may be one of the causes of the observed

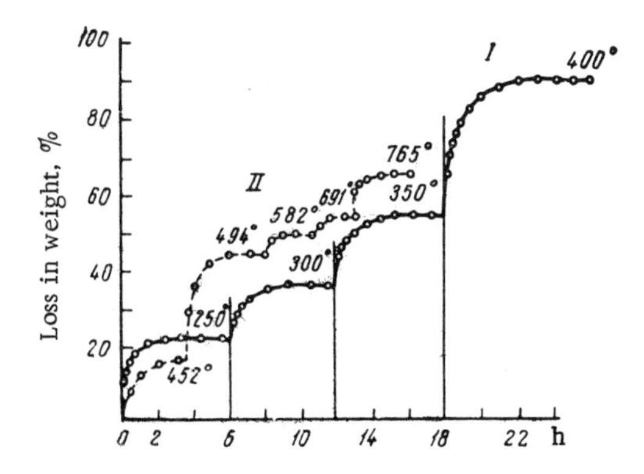

Fig. 1. Curves of the decomposition of polyacrylonitrile (I) and polyoxyphenylene (II) with successive increases in the temperature of treatment.

relationship. A confirmation of this may be the fact that the introduction into polymers with systems of conjugation of groups interfering with their coplanarity and, therefore, with the degree of delocalization of the π-electrons, leads to a decrease in the heat stability of the polymers. This is also shown by the parallel changes of the thermal stability and electrical conductivity of polymers during their thermal treatment.

Another cause of the stabilization and the increase in the thermal stability of polymers may be connected with the increase in the content of highly polymeric fractions which have a structure energetically favorable for the transition of the molecule into the paramagnetic ion-radical state. Paramagnetic macromolecules or "centers of local activation," as they are called, form complexes with the polymer molecules and facilitate the destruction of the radicals arising by bond rupture. The mechanism of the action of such a paramagnetic fraction is, apparently, similar to the mechanism of the inhibition of oxidation processes. It will be shown below that the centers of local activation determine many properties of polymers with conjugated chains.

INFLUENCE OF THE STRUCTURE OF THE MACROMOLECULES ON THE ELECTRICAL PROPERTIES

In an extremely short space of time, synthetic chemists have acquired an extremely broad collection of substances with semiconducting properties. They include substances with a conductivity similar to that of metals, substances which are almost insulators, photosensitive materials, polymers with a high molecular weight, and low-molecular-weight substances which are soluble or insoluble and heat-stable. The incredible abundance and the ever-widening assortment of materials has resulted in an urgent problem for workers in this field — to find the relationship between the electrical properties and the chemical and supermolecular structures of a polymer.

Dulov and Slinkin [39], using polyazophenylene, have shown that the introduction of one methylene group between phenylene groups increases the electrical conductivity of the product, probably because of the denser packing of the molecules. Dense packing of the molecules becomes possible thanks to the flexibility which the $-CH_2-$groups create for them. When two methylene groups are present, the conductivity falls because of the disturbance to the conjugation chain. Similar results on the influence of a disturbance in the conjugation chain on the electrical properties of polymers have also been obtained in the case of the polyazides [40-42], polychelate compounds [41, 42], and polyferrocenes [43].

The supermolecular structure of polymers exerts an influence on their electrical properties which is not less than that of their molecular structure. Airapetyants et al. [44] have studied the influence of preliminary orientation on the electrical properties of products of the thermal conversion of polyacrylonitrile. It was shown by an x-ray study that on thermal treatment the orientation of the molecules as a whole is retained. It was found that this orientation of the molecules leads to a decrease in the specific resistance by approximately an order of magnitude and that the activation energy and the thermo-emf do not depend on the degree of stretching. It is obvious that the resistance falls as a result of the straightening of the molecules with a decrease in the number of intermolecular barriers per unit length of the sample.

It has been shown with polyacetylene [45] that a highly crystalline polymer has a conductivity four orders of magnitude greater than that of an amorphous polymer and an activation energy only half as great. This is connected with the easing of the transfer of carriers from molecule to molecule in crystalline samples.

However, the conclusion drawn from this work [45] cannot be taken literally. High crystallinity must lead to an increase in conductivity only where the coplanarity of the segments of the chains of conjugated double bonds is not disturbed.

In particular, the appearance of crystalline samples in a solid phase may be accompanied by a disturbance in the conjugation between the segments of the molecule because of the departure of the individual segments from coplanarity. This effect has been observed [46] in more than thirty polymers with systems of conjugated bonds and with molecular weights from 1500 to 8000. The polymers investigated belonged to the class of polyazines and polymeric Schiff bases. The clearest influence of crystallinity was found in the polyazines.

The activation energy, electrical conductivity, EPR signal, and even the color of a polymer serve as indications of the greater or smaller degree of conjugation and of the delocalization of the electrons connected with it. Specimens with a greater degree of crystallinity (which was revealed by the sharp reflections in the x-ray diagram) as a rule had a less deep coloration and a greater activation energy and exhibited a lower concentration of unpaired spins per gram. Judging from their molecular weights, the mean lengths of the chains of the polymers studied considerably exceeded the maximum interplanar distances in the polyconjugated systems examined (8-11 Å). Consequently, the molecules of the polymers have individual segments in different crystal cells. A departure from coplanarity always occurs where dense packing of the molecules in the crystal is incompatible with coplanarity.

It must be mentioned that the electrical conductivity, activation energy, and EPR spectrum can by no means always be used as indications of the degree of coplanarity, since this depends on many factors (see Chapter IV). From an analysis of all the properties taken together it follows most frequently that a supermolecular structure that disturbs the coplanarity of the individual blocks displaces the characteristics of a polymer towards those of dielectrics.

INFLUENCE OF PRESSURE ON THE ELECTRICAL AND OPTICAL PROPERTIES OF ORGANIC SEMICONDUCTORS

In addition to the effects of orientation and crystallinity, the influence of pressure on the electrical and optical properties of organic semiconductors has been studied. Generally, the conductivity of organic semiconductors increases with an increase in the pressure (Fig. 2). Various authors [47-49] state that this may be connected with a decrease in the activation energy of the formation of carriers. Thus, for ferrocene the logarithm of the conductivity rises with an increase in the pressure in proportion to the change in volume ΔV [50]. According to other observations [47, 51] the logarithm of the conductivity of some polymers is proportional to the square root of the pressure. Such relationships are not pre-

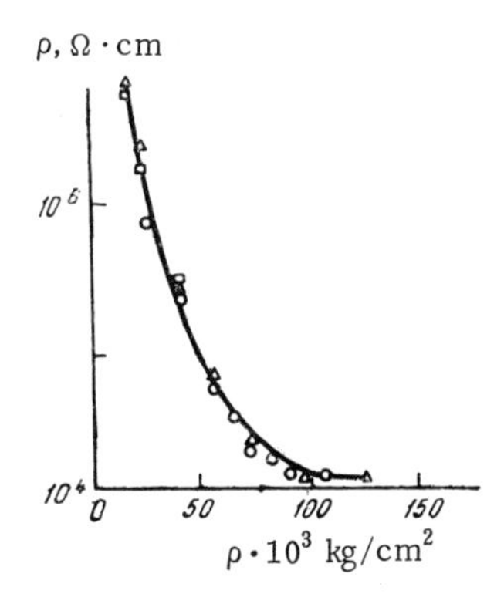

Fig. 2. Specific resistance ρ of the α,α'-diphenyl-β-picrylhydrazyl radical as a function of the pressure P [79].

served over the whole range studied. Generally, the results obtained are interpreted as resulting from a more complete overlapping of the wave functions of adjacent molecules, which leads to a decrease in the energy barriers and to an increase in mobility [52] if a jump mechanism or a change in the band structure of the substance is assumed. Attempts to calculate the change in the energy bands of polycyclic aromatic hydrocarbons and quinones have been made on the basis of approximate considerations of the band structure of anthracene at a pressure of 160 kilobars [54]. On using the structure of the energy bands for the electron in anthracene obtained theoretically by Le Blanc [53], it was found that the widths of the bands increased for holes and electrons by factors of 5.8 and 5.2. In order to show the contribution resulting from the change in mobility and that from the change in concentration of carriers, data for quaterrylene and violanthrone are given (Table 1).

TABLE 1. Influence of Pressure on the Electrical Properties of Quaterrylene and Violanthrone

Substance	Pressure, kg/cm²				n_{160}/n_{atm}	μ_{160}/μ_{atm}
	1		160 000			
	σ, $\Omega^{-1}\cdot cm^{-1}$	ΔE, eV	σ, $\Omega^{-1}\cdot cm^{-1}$	ΔE, eV		
Quaterrylene	$8.3\cdot 10^{-9}$	0.61	$1.4\cdot 10^{-4}$	0.16	$6\cdot 10^{3}$	3
Violanthrone	$4.3\cdot 10^{-11}$	0.78	$1.4\cdot 10^{-3}$	0.20	$8\cdot 10^{4}$	400

It can be seen from these calculations that a decrease in resistance is associated with an increase in the carrier concentration by a factor of 10^3–10^4 with an increase in mobility by a factor of 10^1–10^2.

A somewhat different approach to the calculation of energy bands [55] consists in expressing the activation energy of conduction proper in terms of the ionization potential I, the electron affinity A of the molecule in the gas phase, and the polarization energy P of a crystal with unit charge:

$$\Delta E = I - A - 2P. \tag{1}$$

When a crystal is compressed, the polarization energy P increases, which leads to a change in the activation energy. The change in P can be calculated from the following simple model. Let an ion be placed in a spherical cavity of radius r, when the polarization energy can be written in the form

$$P = \frac{e^2\left(1 - \frac{1}{\varepsilon}\right)}{2r}, \tag{2}$$

where ε is the dielectric constant. A change in the pressure causes the following change in polarization, ΔP:

$$\Delta P = P_2 - P_1 = P_1\left[\frac{r_1}{r_2}\,\frac{\varepsilon_1 - \frac{\varepsilon_1}{\varepsilon_2}}{\varepsilon_1 - 1} - 1\right]. \tag{3}$$

This model has been used to calculate the polarization of a number of substances on the assumption that the compressed substance is isotropic, the increase in conductivity being ascribed only to the change in the activation energy. As can be seen from Table 2, an increase in polarization does in fact lead to a decrease in the activation energy ΔE, which qualitatively agrees with experiment.

In the case of polymeric organic semiconductors, the influence of the pressure reduces mainly to increasing the mobility of the current carriers as a consequence of the facilitation of their jumps between regions of polyconjugation.

TABLE 2. Correlation between the Change in the Polarization of Activation Energy of Conductivity under Pressure

Substance	Pressure, kg/cm^2	$2\Delta P$ (in eV) at $\varepsilon = 3.5$	ΔE, eV
Pentacene	Up to 100 000	0.9	0.75
	0—160 000	1.1	1.36
	0—200 000	1.2	1.4
Quaterrylene	0—160 000	0.9	0.45
Violanthrone	0—160 000	1.1	0.58
Phenylenediamine$_3$ (p-chloranil)$_2$	0—50 000		0.45
	0—300 000		0.49
Metal-free phthalocyanine	0—50 000	0.8	0.37
Copper phthalocyanine	0—50 000	0.8	0.13
Naphthacene	0—100 000	1	0.33
	100—200 000	0.23	0.06
	200 000—300 000	0.03	0.55
Hexacene	0—100 000	1	0.50
Isoviolanthrone	100—300 000	0.26	0.72
Diphenylpicrylhydrazyl	0—100 000		
Perylene-TCE complex	0—100 000	0.9	0.64
N,N,N',N'-Tetramethyl-p-phenylenediamine-p-chloranil complex	0—280 000	1.5	0.24
Perylene-p-chloranil complex	0—12 500	0.29	0.11
1,6-Diaminopyrene-p-chloranil complex	0—100 000	1.3	0.28

In a study of the influence of pressure on the thermo-emf and conductivity of pyrolyzed polyacrylonitrile [60], it was found that the thermo-emf does not depend on the pressure up to 8 kilobars (Fig. 3). The change in the thermo-emf with the temperature has the same nature as the metals. The electrical conductivity rises with an increase in the pressure. The constancy of the thermo-emf shows that the concentration of current carriers at least does not increase with the pressure, otherwise it would change in the way shown in the figure by the broken curve, which was plotted on the assumption that the carriers of one sign are degenerate. The activation energy of conduction decreases from 0.43 to 0.39 eV with a rise in pressure. This change in the activation energy most probably corresponds to a change in the mobility of the carriers which is determined by jumps between regions with delocalized electrons. Within the framework of this model, the nature of the change in the thermo-emf and its temperature dependence shows that current carriers must be in a degenerate state in regions of polyconjugation [60].

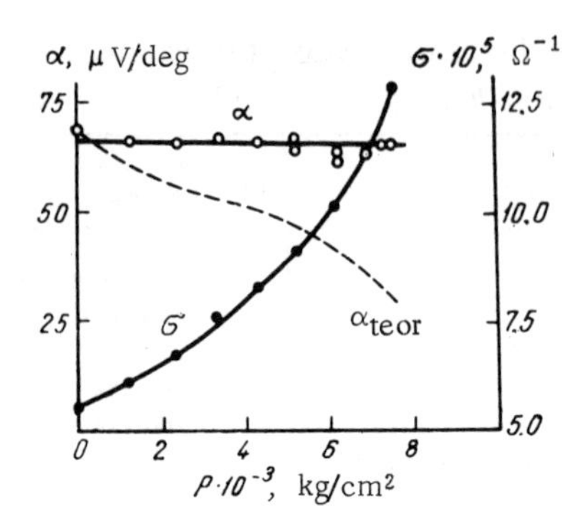

Fig. 3. Electrical conductivity σ and thermo-emf α as functions of the total pressure P for polyacrylonitrile fiber (temperature of treatment 650°C) at room temperature.

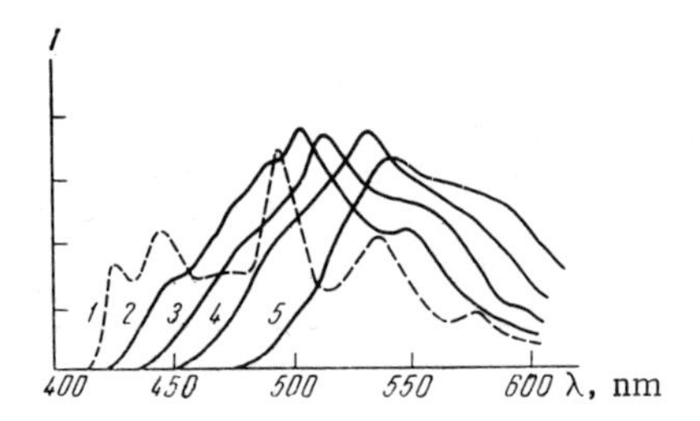

Fig. 4. Shift in the intensity maximum of the fluorescence of anthracene containing tetracene as impurity on the application of different gage pressures: 1) without the application of pressure; 2) 45,000 kg/cm^2; 3) 13,000 kg/cm^2; 4) 26,000 kg/cm^2; 5) 39,000 kg/cm^2.

A study of the optical properties of organic semiconductors on the application of pressure confirms the conclusions drawn in studies of their electrical properties. When pressure is applied, a shift and a change in the intensities of the fluorescence spectra of anthracene with tetracene as an impurity is observed (Fig. 4, [56]). Since the transfer of energy in such systems is connected with dipole-dipole interaction, then, as was to be expected, the ratio of the intensities of the fluorescence at zero pressure and at a given pressure is proportional to the ratio of the lattice constants $(r_0/r)^{-6}$. The influence of pressure on the absorption spectra of anthracene, phenazine, and acridine, dissolved in polystyrene, poly(vinyl acetate), and other plastics has been studied by Offen and Park [57], the observed shifts in the absorption spectra being interpreted within the framework of the MacRae theory of the solvate shift [58, 59]. With a rise in pressure, the interaction of the molecules with the solvent increases and a shift in the absorption band consists of the sum of members taking into account the different pair-interactions of the solute and the solvent:

$$-\Delta\gamma = (A + B)\rho(n^2 - 1)(2n^2 + 1) + C\rho^2 + F\rho^4,$$

where ρ is the density, n is the refractive index of the medium, and A, B, C, and F are constants of various types of interaction of the solvent and the solute.

COMPENSATION EFFECT

The large number of synthetic and thermolyzed polymers with conjugated bonds is characterized by different values of the conductivity (over a range of 20 orders of magnitude) and activation energies of conduction (from ~3 to ~0.01 eV) [1, 62, 63, 66].

Only in individual cases have the specific electrical conductivity σ, the activation energy ΔE, and the photoelectric and magnetic properties been studied in detail. Most frequently, particularly in papers on the synthesis of polymers, the polymers obtained are characterized only by the values of σ and ΔE. Nevertheless, it has proved possible to develop a general approach to understanding the dependence of σ on ΔE for all polymers. The Arrhenius law is obeyed for the conductivity as a function of the temperature:

$$\sigma_T = \sigma_0 \exp[-\Delta E/kT]. \tag{4}$$

The first to draw attention to the linear relationship between $\log \sigma_0$ and ΔE was Eley, in a study of electrical conductivity of amino acids [64] and proteins [77]. In the case of low-molecular-weight organic semiconductors, this relationship was established even earlier by Many et al. [78], who, in a comparison of the results of a study of more than thirty low-molecular-weight organic crystals, observed a correlation between σ_0 and ΔE according to which σ_0 rose with an increase in ΔE. This relationship, as will be shown below, is also observed for a definite group of polymeric semiconductors.

This relationship was confirmed more fully in work [65, 66] on polymers relating to various classes. In a consideration of more than a score of polymers, it was found that they can all be separated into two groups (Fig. 5).

Polymers that are insulators at room temperature may be assigned to the first group. For them, as can be seen from the figure, there is a well-defined compensation effect, i.e., a symbatic change in the pre-exponential factor and the activation energy of conduction.

To the second group may be assigned semiconducting polymers with electrical conductivities of 10^{-5}-10^{-10} $\Omega^{-1}\cdot cm^{-1}$, which do not obey a linear relationship and have large values of σ_0 [67]. On the basis of these data alone, it may be assumed that the mechanisms of conductivity in the two groups of polymers are different. A further study has shown that there is also a relationship between ΔE and σ_0 in the substances belonging to the second group, but in this case it is antibatic [68, 69]. A more detailed study of the relationship between ΔE and σ_0 has been carried out by Gel'fman and Luzan [70]. About 200 polymers were considered. The coeffi-

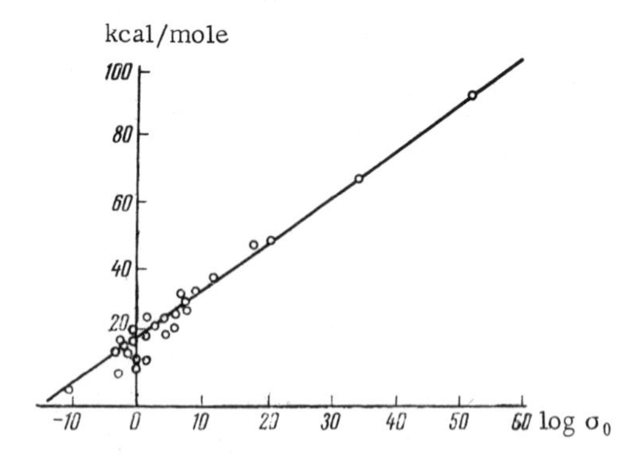

Fig. 5. Relationship between the pre-exponential factor and the activation energy of electrical conduction for a series of polymers.

TABLE 3. Calculation of the Coefficients in Equation (5) for Various Groups of Polymers

Polymers	Correlation coefficient	Level of significance	*b*	*a*	# of pairs n
Substances from paper [66]	0.92	0.999	0.12	0.77	24
Synthetic polymers [71]	0.64	0.999	0.09	1.53	103
Pyrolyzates:					
with an electrical conductivity of $10^{-8}\ \Omega^{-1}\cdot cm^{-1}$ [71]	0.03	0.9	—0.004	0.35	28
irradiated polyethylene [33, 72, 73]	0.47	0.95	—0.075	0.28	13
polyacrylonitrile [8, 74, 75]	0.65	0.95	—0.07	0.36	11
poly(methyl vinyl ketone)[76]	0.82	0.9	—0.21	0.25	5
poly(vinyl alcohol) [71]	0.95	0.95	0.15	1.29	5

cients in the equation

$$\Delta E = a + b \log \sigma_0 \tag{5}$$

were calculated by the method of least squares.

The results are given in Table 3.

It follows from the table that, in agreement with earlier work, all the synthetic polymers are characterized by compensation effects. These polymers dissolve in various solvents and have a short chain of linear continuous conjugation and low conductivities. A relationship between the activation energy of the conductivity and pre-exponential factor which is the reverse of the compensation effect is observed for highly conducting polymeric semiconductors [with the exception of pyrolyzates of poly(vinyl alcohol)] obtained at high temperatures and having a region of three-dimensional polyconjugation in their structure. As can be seen from Table 3, for the pyrolyzates the numerical values of the coefficients are similar to one another regardless of the structure of the initial polymers. It may be concluded from the data given that these two groups, which include the majority of polymers with conjugated bonds obtained up to the present time, differ in their conduction mechanisms. It is interesting that if any impurities that change ΔE and σ_0 are introduced into a polymer, the linear relationship between these quantities is preserved [69].

REFERENCES

1. A.V. Topchiev (Editor), Organic Semiconductors [in Russian], Izd. Akad. Nauk SSSR, Moscow (1963).
2. Y. Okamoto and W. Brenner, Organic Semiconductors, Reinhold, New York (1964).
3. A. A. Berlin, Khim. i Tekhnol. Polimerov, 1960(7-8):139.
4. A. A. Berlin, M. I. Cherkashin, O. G. Sel'skaya, and V. S. Pimakov, Vysokomolek. Soedin., 1:1817 (1959).
5. A. A. Berlin, Khim. Prom., 6:6 (1960).
6. A. A. Berlin, Khim. Prom., 12:23 (1962).
7. B. É. Davydov, B. A. Krentsel', Yu. A. Popov, and L. V. Prokof'eva, Vysokomolek. Soedin., 5:321 (1963).
8. A. V. Topchiev, M. A. Geiderikh, V. A. Kargin, B. A. Krentsel', B. É. Davydov, L. S. Polak, and I. M. Kustanovich, Dokl. Akad. Nauk SSSR, 128:312 (1959).
9. C. S. Marvel and J. H. Rossweiler, J. Am. Chem. Soc., 80:1197 (1958).
10. A. A. Berlin, N. G. Matveeva, and A. I. Sherle, Dokl. Akad. Nauk SSSR, 140:368 (1961).
11. A. V. Topchiev, Yu. V. Korshak, B. É. Davydov, and B. A. Krentsel', Dokl. Akad. Nauk SSSR, 147:645 (1963).
12. F. Winslow, W. Barker, and W. Jager, J. Polymer Sci., 16:101 (1955).
13. F. Winslow, W. Barker, and W. Jager, J. Am. Chem. Soc., 77:4751 (1955).
14. N. Grassie, Chemistry of High Polymer Degradation Processes, Butterworth, London (1958).
15. L. Tokarzewski, Roczniki Chemii, 33:619 (1959).
16. A. A. Berlin, R. M. Aseeva, G. I. Kamchev, and E. L. Frankevich, Dokl. Akad. Nauk SSSR, 144:1042 (1962).
17. N. A. Slovokhotova and I. V. Astaf'ev, Vysokomolek. Soedin., 3:1607 (1961).
18. I. Bohrer, Trans. N.Y. Acad. Sci., 20:367 (1958).
19. H. Mark, Proceedings of a Symposium on the Role of Solid State Phenomena in Electric Circuits, 7:125 (1959).
20. M. Bacher and H. Mark, Angew. Chem., 73:641 (1963).
21. N. Grassie, Trans. Faraday Soc., 48:379 (1952).
22. L. Drechsel and P. Görlich, Jenaer Jahrb., 1:165 (1963).
23. L. Drechsel and P. Görlich, Infrared Physics, 3:229 (1963).
24. H. Pohl, Proceedings of the 4th conference on Carbon, Pergamon Press, New York (1960), p. 241.
25. H. Pohl and I. Laherrere, Ibid., p. 259.
26. C. S. Marvel and H. Hill, J. Am. Chem. Soc., 72:4819 (1950).
27. M. Geiderikh, B. É. Davydov, B. A. Krentsel', I. M. Kustanovich, L. S. Polak, A. V. Topchiev, and R. M. Voitenko, Proceedings of an International Symposium on Macromolecular Chemistry. Section 3 [Russian version], Izd. Akad. Nauk SSSR (1960), p. 85.
28. V. A. Kargin and I. A. Litvinov, Vysokomolek. Soedin., 7:262 (1965).
29. É. A. Silin', D. A. Plumane, and A. V. Airapetyants, Élektrokhimiya, 2:608 (1966).
30. É. A. Silin', V. A. Motorykina, I. K. Shmidt, M. A. Geiderikh, B. É. Davydov, and B. A. Krentsel', Élektrokhimiya, 2:117 (1966).
31. G. Oster, G. Oster, and M. Kryszewsky, Nature (London), 191:164 (1961).

32. W. Bobeth, A. Heger, and A. Weihs, Khim. i Tekhnol. Polimerov, 4:101 (1962).
33. N. A. Bakh, V. D. Vityukov, A. V. Vannikov, and A. D. Grishina, Dokl. Akad. Nauk SSSR, 144:135 (1962).
34. A. V. Vannikov and N. A. Bakh, Élektrokhimya, 1:617 (1965).
35. F. Eirich and G. Mark, Khim. i Tekhnol. Polimerov, 3:16 (1961).
36. W. Skoda, J. Schurz, and H. Bayzer, Khim. i Tekhnol. Polimerov, 9:43 (1960).
37. M. A. Geiderikh, B. É. Davydov, and B. A. Krentsel', Izd. Akad. Nauk SSSR, Ser. Khim., 1965:636.
38. A. A. Berlin, B. I. Liogon'kii, and V. A. Bonsyatskii, Dokl. Akad. Nauk SSSR, 144:1316 (1962).
39. A. A. Dulov and A. A. Slinkin, Dokl. Akad. Nauk SSSR, 143:1355 (1962).
40. A. V. Topchiev, Yu. V. Korshak, B. É. Davydov, and B. A. Krentsel', Dokl. Akad. Nauk SSSR, 147:645 (1963).
41. V. M. Vozzhennikov, Z. V. Zvonkova, E. G. Rukhadze, G. S. Zhdanov, and V. P. Glushkova, Dokl. Akad. Nauk SSSR, 143:1131 (1962).
42. A. P. Terent'ev, V. V. Rode, E. G. Rukhadze, V. M. Vozzhennikov, Z. V. Zvonkova, and L. I. Badzhadze, Dokl. Akad. Nauk SSSR, 140:1043 (1961).
43. A. N. Nesmeyanov, V. V. Korshak, and S. L. Sosin, Dokl. Akad. Nauk SSSR, 137:1370 (1961).
44. A. V. Airapetyants, R. M. Voitenko, B. É. Davydov, B. A. Krentsel', and V. S. Serebryannikova, Vysokomolek. Soedin., 6:86 (1964).
45. M. Hatano, S. Kambara, and S. Okamoto, J. Polymer Sci., 51:156, 526 (1961).
46. B. É. Davydov, R. Z. Zakharyan, G. P. Karpacheva, B. A. Krentsel', G. A. Lapitskii, and G. V. Khutareva, Dokl. Akad. Nauk SSSR, 160:650 (1965).
47. G. A. Samara and H. G. Drickamer, J. Chem. Phys., 37:474 (1962).
48. R. B. Aust, W. H. Bentley, and H. G. Drickamer, J. Chem. Phys., 41:1856 (1964).
49. W. H. Bentley and H. G. Drickamer, J. Chem. Phys., 42:1573 (1965).
50. Y. Okamoto, I. Y. Chang, M. A. Kantor, J. Chem. Phys., 41:4010 (1964).
51. H. A. Pohl, A. Rembaum, and A. Henry, J. Am. Chem. Soc., 84:2699 (1962).
52. D. W. Wood, T. N. Andersen, and H. Eyring, J. Phys. Chem., 70:360 (1966).
53. O. H. LeBlanc, J. Chem. Phys., 35:1275 (1961).
54. Y. Harada, Y. Marujana, I. Shirotani, and H. I. Inokuchi, Bull. Chem. Soc. Japan, 37:1378 (1964).
55. M. Batley and L. E. Lyons, Austral. J. Chem., 19:345 (1966).
56. H. Ohigashi, I. Shirotani, H. Inokuchi, and S. Minomura, J. Phys. Soc. Japan, 19:1996 (1964).
57. H. W. Offen and E. H. Park, J. Chem. Phys., 43:1848 (1965).
58. H. W. Offen, J. Chem. Phys., 42:430 (1965).
59. H. W. Offen, J. Chem. Phys., 42:2523 (1965).
60. A. V. Airapetyants and R. M. Vlasova, Élektrokhimiya, 1:1400 (1965).
61. H. Pohl, Chem. Ing. News, 40:86 (1962).
62. A. A. Dulov, Usp. Khim., 35:1853 (1966).
63. V. V. Pen'kovskii, Usp. Khim., 33:1232 (1964).
64. G. Cardwell and D. Eley, Disc. Faraday Soc., 27:115 (1959).
65. E. I. Balabanov, A. A. Berlin, V. P. Parini, V. L. Tal'roze, E. L. Frankevich, and M. I. Cherkashin, Dokl. Akad. Nauk SSSR, 134:1123 (1960).

66. V. L. Tal'roze and L. A. Blyumenfel'd, Dokl. Akad. Nauk SSSR, 135:1450 (1960).
67. V. L. Tal'roze and L. A. Blyumenfel'd, Vysokomolek. Soedin., 4:1282 (1962).
68. A. V. Airapetyants, R. M. Voitenko, B. É. Davydov, and V. S. Serebryanikov, Vysokomolek. Soedin., 3:1876 (1961).
69. N. A. Bakh, A. V. Vannikov, A. D. Grishina, and S. V. Nizhnii, Usp. Khim., 34:1733 (1965).
70. A. Ya. Gel'fman and R. G. Luzan, Dokl. Akad. Nauk SSSR, 168:1371 (1966).
71. A. Ya. Gel'fman, Technical and Economic Information. Production of Chemical Reagents. Electrophysical Properties of Polymers [in Russian], No. 2 (8), 7 (1965).
72. A. V. Vannikov and N. A. Bakh, Dokl. Akad. Nauk SSSR, 149:357 (1963).
73. A. V. Vannikov, Dokl. Akad. Nauk SSSR, 152:905 (1963).
74. V. A. Kargin, A. V. Topchiev, B. A. Krentsel', L. S. Polak, and B. É. Davydov, Zh. Vses. Khim. Obshchestva im. D. I. Mendeleeva, 5:507 (1960).
75. R. M. Voitenko and É. M. Raskina, Dokl. Akad. Nauk SSSR, 136:1137 (1961).
76. A. A. Dulov, A. A. Slinkin, and A. M. Rubinshtein, Izv. Akad. Nauk SSSR, Ser. Khim., 1964:26.
77. D. Eley and D. Spivey, Trans. Faraday Soc., 156:1432 (1960).
78. A. Many, E. Harnik, and D. Gerlich, J. Chem. Phys., 23:1733 (1955).
79. H. Inokuchi, I. Shirotani, and S. Minomura, Bull. Chem. Soc. Japan, 37:1234 (1964).

CHAPTER II

THE PHOTOCONDUCTIVITY OF POLYMERIC SEMICONDUCTORS

A study has been made of the spectral characteristics and photoelectric sensitivity of many organic polymers with conjugated double bonds. In spite of the considerable number of papers devoted to photoelectric sensitivity, so far there has been no sufficiently well-founded theory of the mechanism of the photoexcitation of the current carriers and their transfer in the polymeric structure. Nevertheless, the characteristics of the photoconductivity of several polymers with conjugated bonds give grounds for considering these polymers as promising materials for practical use.

There are no experimental results available for polymeric semiconductors that could sufficiently fully explain the processes of photoconduction. Such results have been obtained for low-molecular-weight semiconductors [1-3].

As a result of the study of these latter compounds, clear ideas have been developed on the mechanism of the photoexcitation of free carriers. The absorption of light by a crystal leads to the formation of molecular singlet excitons. The effect of the change of photoconductivity in a magnetic field makes it possible to study experimentally the transfer excitons that are formed from the molecular excitons and which directly precede the appearance of the free current carriers [4-6]. Numerous studies have been devoted to the mechanism of the movement of the photoexcited free carriers in low-molecular-weight organic crystals (cf. Chapter IV). The lifetime of the carriers and the characteristics of the local levels upon which the drift mobility and the photoelectric sensitivity itself depend have been determined. The influence of impurities introduced into the crystals in rigidly defined amounts on their photosensitivity and, in particular, on the processes of recombination of the current carriers has been considered. There are no similar investigations of these parameters for polymeric semiconductors. As a rule, investigations have been limited to studying the electronic absorption spectra, the spectral distribution of the stationary photoelectric sensitivity, and its temperature dependence.

In the majority of papers, data are given on the stationary photoconductivity in a constant electric field. For the reliable measurement of the photoelectric sensitivity, the photoconductivity σ_{ph} must be at least an order of magnitude greater than the dark conductivity σ_{dk}. The low quantum yield, the short lives of the carriers, and the structural nonuniformity of the substances considered does not permit the expectation of large values of σ_{ph}. Consequently, investigations are generally carried out on high-ohmic semiconducting polymers. If these polymers have been obtained by thermal treatment, their low conductivity indicates considerable segments with mainly ordinary C—C bonds which obstruct the transfer of the carriers from one region to another. The question immediately arises of whether the absorption of light and the formation of free current-carriers takes place in the regions of polyconjugated double bonds or in the segments with saturated bonds.

ABSORPTION SPECTRA OF POLYMERIC SEMICONDUCTORS

It has been shown [7, 8] that the absorption of light takes place in the regions of polyconjugation. When polyacrylonitrile was subjected to thermal treatment, two forms of the transformation of the polymer were observed. At low temperatures (up to 300°C) the number of sections of conjugated bonds of a definite length rose. In the spectrum an increase in the intensity of the absorption in the 350 nm region with no change in the shape of the absorption spectra was seen, as is shown in Fig. 6. When the temperature of treatment was raised, the cross-linking of the sections of continuous conjugation took place, which is shown in a rise in the absorption in the long-wave fall of the curve (450–600 nm).

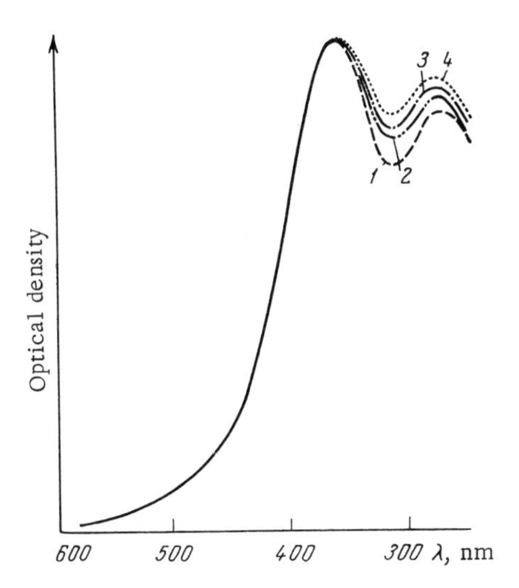

Fig. 6. Spectral curves of the optical absorption of polyacrylonitrile at various stages of thermal treatment at 220°C: 1) 1 h; 2) 2 h; 3) 3 h; 4) 4 h. Curves reduced to the same value of the optical density at the maximum at 355 mμ.

In this case there was a symbatic displacement of the fall into the long-wave region and growth of the regions of polyconjugation.

This relationship is found more fully in a study of the electronic absorption spectra of polymeric semiconductors obtained by the radiation-thermal modification of polyethylene [9, 10], since the increase in the size of the segments with polyconjugated bonds has been confirmed by the EPR method and by alternating-current measurements. The electronic absorption spectra of the materials are characterized by strong absorption in the short-wave region ($k \gg 10^4$ cm^{-1}), a fairly sharp boundary of the long-wave absorption, and relatively weak absorption in the long-wave region ($k < 10^2$ cm^{-1}). As the temperature of preparation of the samples is increased, the boundary of the long-wave absorption shifts from the visible region into the infrared. In addition to this, the absorption in the long-wave region increases. At temperatures of treatment > 700°C, it becomes so large that the boundary of the long-wave absorption disappears.

A study of the electron paramagnetic resonance spectra and the electrical characteristics of samples with increasing temperature of treatment shows an increase in size of the polyconjugated regions. At the same time, as is well known (see, for example, [11]), with an increase in the size of the system of polyconjugated bonds the electronic absorption shifts into the long-wave region of the spectrum. Thus, the displacement of the spectra into the long-wave region, which is symbatic with the increase in the size of the regions of polyconjugation, as the temperature of treatment is increased shows that the observed electron density is connected with the regions of polyconjugation and is determined by the energy structure of the electronic system of these regions. Similar conclusions have been arrived at by a number of other authors [12-14].

The electronic absorption spectra have one interesting feature. As the conductivity of the polymers is increased (which is generally brought about by increasing the temperature of treatment), the absorption rises very strongly in all regions of the spectrum up to $\sim 15\ \mu$, including the visible region of the spectrum. At the same time, the long-wave fall is replaced by uniform absorption within the spectral limits mentioned [12, 13]. It has been shown [14] for the initial stage of this process, using complexes of polytetracyanoethylene with metals, that the long-wave fall of the electronic absorption spectra is extremely diffuse. Since this fall is determined by the dimensions of the regions of polyconjugation, in the opinion of the authors this diffuseness of the edge of the absorption band shows that there is a range of regions of polyconjugation of different lengths. This "diffuseness" is the more strongly expressed the higher the conductivity of the polymer.

However, another interpretation of this phenomenon can be given. According to various estimates, in highly conductive polymeric semiconductors a high concentration of free charge-carriers ($\sim 10^{19}$ cm^{-3}) should be observed. It is also known that in inorganic semiconductors the absorption of free carriers accompanied by intraband transitions is observed in just this region of the spectrum. Since the approximation of the band theory for highly conducting polymeric semiconductors has been fully substantiated, the observed absorption may be ascribed to the free carriers. Both these assumptions are quite probable. A distribution of the macromolecules with respect to their molecular weights [15] favors the first assumption, while it is known that the lengthening of the conjugating system does not lead to a symbatic shift of absorption into the long-wave region. On the contrary, the latter tends to a definite limit [10, 16]. This is, rather, in agreement with the second hypothesis. However, it is most likely that both processes occur. A definitive interpretation of the observed effect requires additional confirmation.

Thus, it may be considered as established that the change in the absorption spectra unambiguously characterizes an increase in the size of the regions of polyconjugated bonds.

The least well explained are those processes that take place after the absorption of a photon and lead to the appearance of current carriers. The fact that photoconductivity has a purely electronic and not an ionic nature in organic polymers has been demonstrated [17] in a study of the relaxation of the photoconduction in the range from 10^{-5} to 10^{-2} sec using a polymer obtained by the oxidative dehydropolycondensation of p-diethynylbenzene and phenylacetylene. An analysis of the relaxation of the photocurrent has enabled a rapid component ($\sim 10^{-5}$ sec) to be isolated and has shown that the relaxation is well described by a hyperbolic relationship, i.e., has a bimolecular character.

The majority of investigations is limited to a phenomenological description of the processes taking place in the irradiation of polymeric semiconductors.

PHOTOCONDUCTIVITY OF POLYMERIC SEMICONDUCTORS

Photoelectric sensitivity has been detected in a large number of polymeric semiconductors [18-37]. A detailed study has been carried out on samples of radiation-thermally modified polyethylene [9]. Photoelectric sensitivity was detected over the whole of the region of the spectrum investigated from the ultraviolet to the near infrared (300-900 nm). The magnitude of the observed photocurrent was determined by the wavelength of the incident light and, for example, for $\lambda = 600$ nm at room temperature it was between 2×10^{-14} and 1×10^{-12} A for various samples with a superimposed sample voltage of 400 V and an energy of the incident radiation of ~ 6 mW/cm^2. For all the samples, $\sigma_{ph}/\sigma_{dk} > 10$.

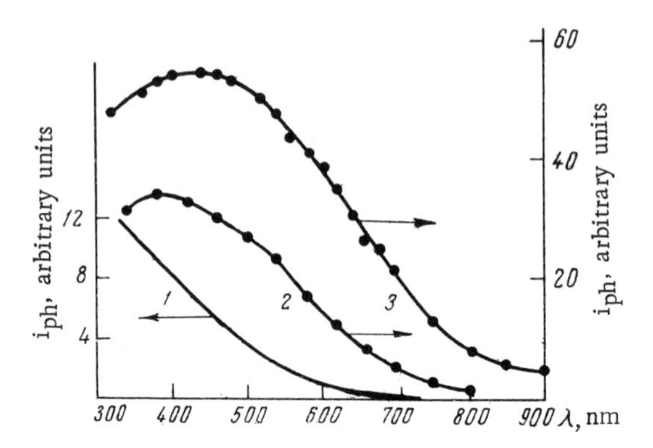

Fig. 7. Spectral characteristics of the photoelectric sensitivity of modified polyethylene. Temperature of treatment: 1) 265°C; 2) 360°C; 3) 430°C.

A general tendency in the intensification of the thermal treatment of the samples is a rise in their photoconductivity and their dark conductivity. These two magnitudes rise to approximately equal extents, so that the factor for the change in conductivity on illumination remains constant. At a voltage of up to 400 V, Ohm's law remains valid for the photocurrent and within a wide range of illuminations the photocurrent depends linearly on the intensity of the incident light.

Figure 7 gives the spectral characteristics of the photoelectric sensitivity. With a rise in the temperature of treatment, the maximum of the photoelectric sensitivity shifts in the long-wave direction in a similar manner to that observed for the electronic absorption spectra. With a rise in the temperature, the photoconductivity changes exponentially. The activation energies of photoconduction for various wavelengths of the incident light for three samples studied are given in Table 4.

The following conclusions can be drawn from a consideration of the table: in the first place, when the temperature of treating the sample is raised, there is a decrease in the activation energy of both the photoconduction and the dark conduction if the wavelength of the exciting light is constant; and, in the second place, for samples obtained at one and the same temperature of treatment there is a decrease in the activation energy of the photoconduction with a decrease in the wavelength of the incident light.

TABLE 4. Activation Energy of the Photocurrent (ΔE_{ph}, eV) for Various Wavelengths

Temperature of treatment of the sample, °C	ΔE_{dk} eV	λ, nm				
		800	600	546	500	400
260	0.60		0.19	0.175	0.158	0.150
350	0.50		0.073	0.095		
430	0.40	0.075		0.071		0.065

A study of polyphenylenediacetylene in which all the atoms of hydrogen of the main chain havc been replaced by phenyl radicals has shown that the photoelectric sensitivity is connected with the presence of a sufficiently extended system of conjugated bonds [23]. As in the case of radiation-thermally modified polyethylene, it was found that the activation energy of the photoconduction is considerably smaller than that of the dark conduction (1.67 eV) and rises with the wavelength from 0.15 eV (360 nm) to 0.28 eV (800 nm). The authors consider that the activation energy of the photoconduction is equal to the thermal energy of the transfer of the excitation of the carriers into the conduction band from the trapping levels into which they fall after the illumination of the sample. Similar results have been obtained by Demidova et al. [20].

Since photoconductivity is observed in the region of the absorption of the conjugated bonds, its changes reflect transformations in the regions of continuous conjugation which determine the electrophysical properties of the polymer. By comparing the photo and dark properties it is possible to obtain additional information on the processes of conduction. As already mentioned, when polyacrylonitrile is subjected to thermal treatment at temperatures above 300°C, the absorption in the long-wave region begins to rise and there is a simultaneous increase in the photoconductivity [7, 8]. However, the increase in photoconductivity cannot be connected solely with the increase in absorption; if absorption increases by a factor of 3, photoconductivity rises by a factor of 15, i.e., 5-7 times greater. The dark conductivity rises by the same factor. The increase in the photoconductivity and dark conductivity have the same cause — an increase in mobility, since this comes in as a common factor in the expression for the photo- and dark conductivities.

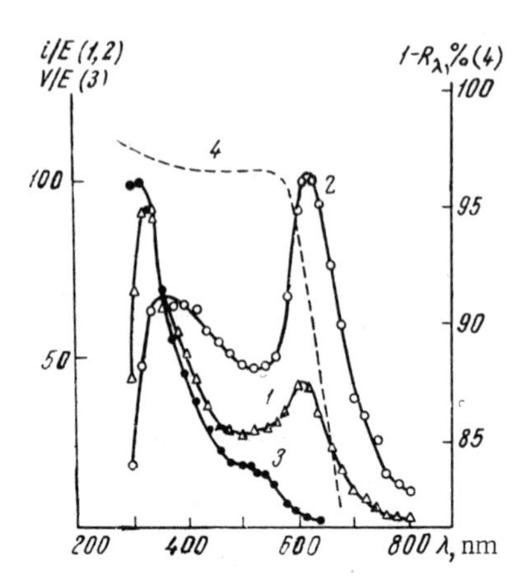

Fig. 8. Spectral characteristics of poly-p,p'-diethynylazobenzene. 1) Photoconductivity; 2) photoconductivity after previous illumination of the layer with UV light; 3) photo-emf; 4) absorption spectrum.

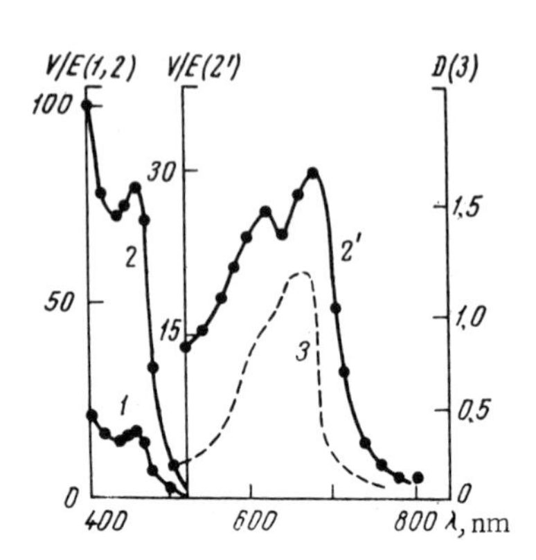

Fig. 9. Spectrum of the photo-emf of copper polyphenylacetylide. 1) Before dyeing; 2, 2') after dyeing in an ethanolic solution of methylene blue; 3) absorption spectrum of an ethanolic solution of methylene blue (10^{-3}M).

CONNECTION BETWEEN OPTICAL AND PHOTOELECTRIC PROPERTIES

When the absorption spectra and photoconductivity spectra for the majority of polymers are compared, a direct correspondence between them can be found. At the same time, by analogy with low-molecular-weight organic semiconductors, it may be assumed that light causes singlet–singlet transitions. In this case, the absorption of a photon in a semiconducting polymer leads to the formation of an exciton which can migrate freely within the limits of a region of conjugated bonds. However, a case is known in which the photoconductivity spectrum does not repeat the form of the absorption spectrum but has the shape typical for inorganic semiconductors [36]. As shown in Fig. 8, in a study of poly-p,p'-diethynylazobenzene a sharp fall in the intensity of absorption in the region of the maximum photoconductivity (610 nm) is observed, which is common for inorganic but unusual for polymeric semiconductors in which, as a rule, the maximum photoelectric sensitivity is in the region of strong absorption. The presence of the photoconductivity maximum on the falling part of the absorption curve is due to the ratio and nature of the surface and bulk recombination.

The subsequent fate of the exciton can be followed by studying a phenomenon connected with the activation of photoconduction by ultraviolet light [36]. It is found that the preliminary irradiation of the polymers with ultraviolet light greatly increases their photoconductivity, as can be seen from Fig. 8. The following explanation of the observed influence of UV irradiation can be given. Ultraviolet light can ionize conjugated molecules, liberating photoelectrons, which are retained in the structure of the polymer, and creating positively charged local centers which can serve as electron traps. The existence of this process is confirmed by a study of the EPR spectra of polymers irradiated with ultraviolet light [37]. On long-wave irradiation, excitons are produced which are destroyed on defects (including defects created by the UV irradiation) with the formation of a trapped electron and a mobile hole. In contrast to the polymers investigated, no characteristic increase in the photoelectric sensitivity after preliminary ultraviolet illumination is found for monomers [22].

INFLUENCE OF THE STRUCTURE OF THE POLYMERS ON THEIR PHOTOCONDUCTIVITY

In considering the mechanism of transfer in the process of photoconduction, two factors may be isolated: the motion of the charge carriers by a band mechanism and the influence exerted on the transfer process, and consequently on photoconductivity, by the dimensions of the system of continuous polyconjugation.

The change in the mechanism of the electronic processes on passing from a solution of a polymer to the solid state has been considered by Myl'nikov [21].

In a study of the copolymer of p-diethynylbenzene and p-nitrophenylacetylene, the absorption spectra were measured on the polymer in solution and the polymer prepared in the form of a solid film. In comparison with the solution, the absorption bands of the individual molecules in the film were highly diffuse. The marked rise in the absorption in the short-wave region of the spectrum shows that the film contains a generalized system of electrons participating in absorption. The photoconductivity spectrum basically reproduces the absorption spectrum of the solid film. On the basis of what has been said, it may be concluded that in the solid polymer energy bands for electrons are realized which cover all the molecules and that, consequently, within certain limits, it is possible to apply the band scheme to this polymer.

This conclusion is apparently limited in nature. In actual fact, in the example given it is possible to assume some similarity of the polymer to the molecular crystals with very narrow bands distributed throughout the samples, which excludes the separation of the total transfer process into movement within the same molecule and charge transfer between molecules. The molecule considered is soluble and, consequently, its molecular weight is insufficient for it to be possible to assume the formation of a band in a single molecule. However, the majority of polymers studied have fairly large systems of continuous conjugation and for these it is possible to separate transfer into the two processes mentioned.

The question of how the photoelectric characteristics change with an increase in the number of conjugated double bonds and, in particular, on passing from monomers to polymers, was apparently posed for the first time in 1963 [22] when the photoconducting properties of metal acetylides were studied.

The compounds investigated, in spite of the small extent of the system of conjugated bonds, possess photoconductivities and photo-emf's exceeding by several orders of magnitude those observed in polymers of the class of the polyacetylenes. The authors concerned consider that the determining factor for phototransfer is formed by the intermolecular configurational interaction, which lowers the potential barriers between the molecules. These interactions can occur by bridges of π-complexes of electron-accepting copper atoms with acetylenic bonds functioning as electron donors. However, this is apparently an unjustifiably simplified approach to the observed situation since it is not so much that the π-complexes are bridges between the molecules as that they modify the system of levels, liberating new mechanisms for the generation of carriers.

The question of the influence of the extent of the system of conjugated bonds on the photoelectric properties of polymers was considered later [20]. In this case, it was not monomers and polymers that were compared but polymers with different molecular weights. For these all the relationships found for other polymers are retained: the linearity of the lux–ampere characteristics, the observance of Ohm's law for photocurrents in fields of up to 5×10^4 V/cm, the dependence of ΔE_{ph} on the wavelength, etc. In addition, it was found that the photocurrent, referred to the same number of incident quanta, falls on passing to compounds with decreasing length of the chain of conjugation.

PHOTOSENSITIZATION EFFECT

The spectral sensitization by dyes of the photo effect, previously found in the study of inorganic semiconductors, is of great importance for understanding the mechanism of photoconduction and also for the practical use of photosensitive materials [38]. It has been possible to observe the spectral sensitization of the photo effect in polymeric organic semiconductors, in particular polymers with triple bonds and polyacetylides [18, 19, 37]. Figure 9 gives the spectral distribution of the photo-emf of copper phenylacetylide before and after dyeing with methylene blue. In the presence of the dye, the measured photo-emf rises sharply (by a factor of 8-10) in the region in which it was observed in the initial polymer and it also rises in the spectral region corresponding to the absorption of the dye. At the same time, the photo-emf spectrum is similar to the absorption spectrum, which shows that the sensitizing action is exerted by the monomolecular form of the dye. The presence of a maximum in the 620 nm region shows that the aggregated form of the dye may also take part in the sensitization of the photo effect. In a similar manner to the photo-emf, sensitization of the photoconductivity of copper phenylacetylides is observed on the adsorption of methylene blue. The features mentioned are repeated in the case of other polymer–dye pairs.

Two explanations exist for the sensitization effect. One connects this effect with the transfer of excitation energy from the adsorbed molecules of the dye to the carriers trapped in the local levels, and their excitation into the conducting state. According to the other model, the process of sensitization consists in the transfer of an electron from the dye to the semiconductor. Since some of the dyes studied do not possess photoelectric sensitivity, the first mechanism is preferable. Because sensitized photoconduction is carried out by holes, the transfer of excitation energy to the polymer apparently leads to the ejection of an electron from a completely filled band into local levels with the liberation of holes taking part in photoconduction or (which is the same thing) to the excitation of holes into the valence band from the trapping levels.

The observed change in the intrinsic photoelectric sensitivity of a polymer in the presence of a dye is not yet completely clear and is possibly connected with the nature of the traps created by the adsorbed molecules of the dye.

The increase in the photoconductivity of polynaphthalene on the addition of traces of iodine may be regarded as a peculiar sensitization phenomenon [31]. In this case the influence of the iodine may be exerted by the second mechanism.

The effect of the increase in photoconductivity of a heterophase system of a polymer with conjugated bonds in the main chain and a low-molecular-weight acceptor, which is closely similar to the phenomenon of sensitization, is of great interest [32]. In this system an increase in photoconductivity by a factor of 10^3–10^8 in comparison with the photoconductivity of the polymer and acceptor taken separately is found. Thus, for the polyphenylacetylene–chloranil system the photoconductivity σ_{ph} is between 0.6×10^{-7} and 2.3×10^{-7} ($\lambda = 440$ nm) as compared with $\sigma_{ph} < 10^{-13}\ \Omega^{-1} \cdot cm^{-1}$ for the polymer without the acceptor. It is important to note that the spectral sensitivity of the photoconductivity in the long-wave region is determined by the absorption spectrum of the polymer. The effect of a considerable increase in photoconductivity when a polymer–acceptor heterophase system is created is observed with a large number of polymers having conjugated bonds.

The difference of the phenomenon considered from the sensitization effect consists in the fact that the increase in photoconductivity in the heterophase system takes place in the region of absorption of the polymer, while on sensitization the photoconductivity of the system increases mainly in the region of the absorption of the dye molecules. In addition, the magnitude of the sensitized photoconductivity does not exceed the photoconductivity of the polymer in the region of maximum absorption, while in the systems considered it exceeds the photoconductivity of the polymer and the acceptor by several orders of magnitude. A possible ex-

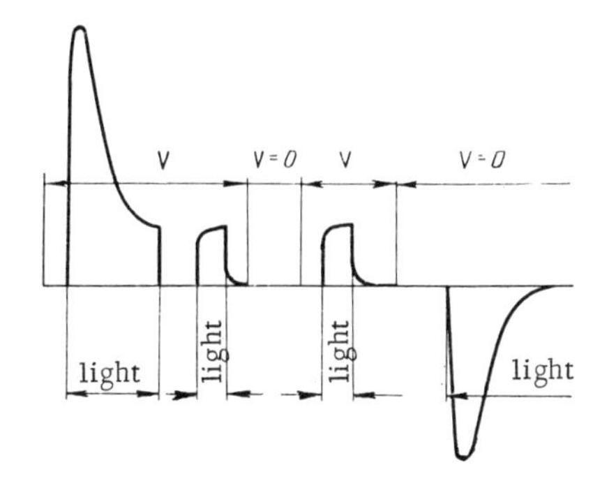

Fig. 10. Kinetics of the photocurrent in a polymeric Schiff's base.

planation of these facts is an increase in the probability of the destruction of an exciton at the polymer–acceptor boundary with the trapping of the electron by an acceptor molecule and the formation of a free hole, which is illustrated by a considerable increase in the quantum yield.

Up to the present time we have considered semiconducting polymers in which the photoconductivity undoubtedly is electronic in nature. It must be noted that in many polymers there is an inertial internal photo effect with a relaxation time $> 10^{-1}$ sec. Figure 10 shows the photocurrent as a function of the time for a polymer belonging to the class of polymeric Schiff's bases [33]. In the first period of illumination, the photocurrent passes through a maximum and then reaches a stationary value. In repeated periods of illumination the stationary photocurrent increases monotonically. When a sample is illuminated in the absence of a field, a depolarization photocurrent is observed. The maximum amplitude for this current i_0 depends on the length of the dark pause. Plotting this relationship as a graph of $1/i_0$ versus τ shows that it obeys a hyperbolic law. Consequently, the "dissipation" of the states of polarization in the dark is bimolecular in nature. Great inertia of the photocurrent and polarization phenomena have also been observed for a number of other polymers [34, 35]. In this case it is not easy to delimit the purely photoelectric effects from photochemical processes and therefore the interpretation of the observed relationships is difficult.

REFERENCES

1. A. N. Terenin, Zh. Vses. Khim. Obshchestva im. D. I. Mendeleeva, 5:498 (1960).
2. A. V. Topchiev, Organic Semiconductors [in Russian], Izd. Akad. Nauk SSSR, Moscow (1963).
3. H. Inokuchi and H. Akamatu, The Electrical Conductivity of Organic Semiconductors [Russian translation], IL, Moscow (1963).
4. E. L. Frankevich and E. I. Balabanov, Zh. Éksperim. Teoret. Fiz., Letters to the Editor, 1(6):33 (1965).
5. E. L. Frankevich and E. I. Balabanov, Fiz. Tverd. Tela, 8:855 (1966).
6. E. L. Frankevich, Zh. Éksperim. Teoret. Fiz., 50:1226 (1966).
7. I. A. Drabkin, L. D. Rozenshtein, M. A. Geiderikh, and B. É. Davydov, Dokl. Akad. Nauk SSSR, 154:197 (1964).
8. I. A. Drabkin and L. D. Rozenshtein, Izv. Akad. Nauk SSSR, Otd. Khim. Nauk, 1964:1113.
9. A. V. Vannikov, G. N. Demidova, L. D. Rozenshtein, and N. A. Bakh, Dokl. Akad. Nauk SSSR, 160:635 (1965).
10. N. A. Bakh, A. V. Vannikov, A. D. Grishina, S. V. Nizhnii, and Z. A. Markova, in: The Radiation Chemistry of Polymers [in Russian], "Nauka," Moscow (1966).
11. J. Murrell, The Theory of the Electronic Spectra of Organic Molecules, Methuen, London (1963).
12. Yu. Sh. Moshkovskii, N. D. Kostrova, and A. A. Berlin, Vysokomolek. Soedin., 3:1669 (1961).
13. Z. A. Markova, B. G. Ershov, and N. A. Bakh, Vysokomolek. Soedin., 6:121 (1964).
14. L. I. Boguslavskii and L. D. Rozenshtein, Élektrokhimiya, 1:713 (1965).
15. A. A. Berlin, Khim. Prom., 1962(12):23.
16. I. A. Misurkin and A. A. Ovchinnikov, Zh. Éksperim. Teoret. Fiz., 4:248 (1966).
17. V. S. Myl'nikov, Dokl. Akad. Nauk SSSR, 148:620 (1963).
18. C. Mylnikov and A. Terenin, Molec. Phys., 2:3871 (1964).
19. V. S. Myl'nikov and A. N. Terenin, Dokl. Akad. Nauk SSSR, 158:1167 (1964).
20. G. N. Demidova, R. N. Pirtskhelava, L. D. Rozenshtein, M. P. Terpugova, and I. L. Kotlyarevskii, Élektrokhimiya, 1:1145 (1965).
21. V. S. Myl'nikov, Dokl. Akad. Nauk SSSR, 157:1184 (1954).
22. V. S. Myl'nikov and A. N. Terenin, Dokl. Akad. Nauk SSSR, 153:1381 (1963).
23. B. É. Davydov, G. N. Demidova, F. M. Nasirov, R. N. Pirtskhelava, and L. D. Rozenshtein, Élektrokhimiya, 1:876 (1965).
24. A. A. Slinkin, A. A. Dulov, and A. M. Rubinshtein, Izv. Akad. Nauk SSSR, Ser. Khim., 1964:1763.
25. H. A. Pohl and D. A. Opp, J. Phys. Chem., 66:2121 (1962).
26. I. L. Kotlyarevskii, L. B. Fischer, A. A. Dulov, A. A. Slinkin, and A. M. Rubinshtein, Vysokomolek. Soedin., 4:174 (1962).
27. L. Drechsel and P. Görlich, Jenaer Jahrb., 1:165 (1963).
28. L. Drechsel and P. Görlich, Infrared Phys., 3:229 (1963).
29. V. S. Myl'nikov, A. M. Sladkov, Yu. P. Kudryavtsev, L. K. Luneva, V. V. Korshak, and A. N. Terenin, Dokl. Akad. Nauk SSSR, 144:840 (1962).

30. G. Oster, G. Oster, and M. Kryszewsky, Nature (London), 191:164 (1961).
31. A. Bradley and J. Hammes, J. Electrochem. Soc., 110:543 (1963).
32. A. A. Berlin, I. A. Drabkin, L. D. Rozenshtein, M. I. Cherkashin, M. G. Chauçer, and P. P. Kislitsa, Izv. Akad. Nauk SSSR, Ser. Khim., 1967:1339.
33. B. É. Davydov, Yu. A. Popov, L. V. Prokof'eva, and L. D. Rozenshtein, Izv. Akad. Nauk SSSR, Otd. Khim. Nauk, 1963:759.
34. B. É. Davydov, I. A. Drabkin, Yu. V. Korshak, and L. D. Rozenshtein, Izv. Akad. Nauk SSSR, Ser. Khim., 1963:1164.
35. L. S. Polak, A. M. Nemchushkin, A. F. Lunin, and Ya. M. Paushkin, in: Semiconducting Polymers with Conjugated Bonds [in Russian], Izd. TsNIITÉneftekhim, (1966), p. 147.
36. V. S. Myl'nikov, E. K. Putseiko, and A. N. Terenin, Dokl. Akad. Nauk SSSR, 149:897 (1963).
37. V. S. Myl'nikov, in: Elementary Photo Processes in Molecules, A. N. Terenin (ed.), "Nauka," Moscow (1966), p. 417.
38. E. K. Putseiko and A. N. Terenin, Zh. Fiz. Khim., 23:676 (1949).

CHAPTER III

CONNECTION BETWEEN THE ELECTRICAL AND PARAMAGNETIC CHARACTERISTICS OF POLYMERIC SEMICONDUCTORS

In the study of the electrical characteristics of the polymeric semiconductors, considerable difficulties are usually encountered since the majority of polymers (particularly synthetic ones) are obtained only in pulverulent form. At the present time it is practically impossible to obtain polycrystalline polymers (and, all the more, monocrystals) since polymers with conjugated bonds usually contain an amorphous phase. With the exception of some individual cases, which will be mentioned below, the Hall effect in these substances cannot be measured. A very general consideration of the structure of polymers with conjugated bonds leads to the conclusion that it is extremely complicated, since it is characterized by the presence of molecules (regions) with mobile π-electrons and contains either intermolecular spaces in which dispersion forces act or sections of saturated covalent bonds. This shows yet again the great complexity of the processes of transfer of the current carriers.

At the present time the unambiguous composition, and all the more the structure, of the substances obtained is unknown not only in pyro polymers but also in polycondensed systems with a considerable degree of purity. Consequently, it is difficult to draw well-founded conclusions on the mechanism of the electronic phenomena observed in semiconducting polymers. In addition to this, almost all the substances studied are characterized by considerable paramagnetism. If it is possible to establish a connection between the paramagnetic and electrical characteristics of these substances, the EPR method will become one of the most powerful methods of studying the electrophysical properties of organic semiconductors.

In principle, it may be expected that three most general groups of electronic states of polymeric semiconductors determining their electric characteristics either directly or indirectly will be responsible for the EPR signal.

The smallest connection with the electrical characteristics will be observed where the EPR spectra record the free radicals trapped in the polymer matrix in the process of synthesis or formed in the cleavage of bonds in the pyrolysis of the polymers [1-3]. In particular, as has been suggested by Berlin [4], it is possible to represent paramagnetic particles as functions of higher-molecular-weight homologs existing in the form of stable double radicals arising in the course of synthesis or as the result of some energetic interaction.

However, in this case it is again possible to trace the influence of the paramagnetic centers on the electrical characteristics since the free radicals associated with the system of conjugated double bonds are characterized by an increased electron affinity and a reduced ionization potential and, because of this, are fairly effective traps for holes and electrons, as has been shown by Frankevich and Tal'roze [5]. Consequently, such states may apparently be effective centers of recombination for electron–hole pairs. In addition to this, showing the imperfection of the structure of the polymer, free radicals may be centers upon which the destruction of the excitons takes place with the formation of free carriers on photoexcitation. Here, however, it is necessary to bear in mind the observed paramagnetism of states with charge transfer in the local centers formed in the polymer [6]. As will be shown below, the presence of two states (ionic and neutral) with similar equilibrium energies and sharply differing intermolecular distances must lead to the appearance of a large number of paramagnetic charged defects. The observed paramagnetism is connected with impurity centers and has no direct relationship with the electrical characteristics.

A second group of states can be detected in those polymeric semiconductors in which the paramagnetic centers are directly connected with donor or acceptor groups. In this case, a direct connection is established between the concentrations of paramagnetic centers and the charge carriers. The establishment of such a relationship enables us to obtain additional information on the mechanism of the generation of the current carriers and the characteristics of their transfer.

The third group of states is observed in substances in which free charge carriers are responsible for the paramagnetism. Here the connection between the two phenomena considered is closest. The properties of the charge carriers appear in the EPR spectra.

The observed electron paramagnetic resonance spectra in polymers with conjugated double bonds have much in common with one another. Because of its pronounced delocalization over the system of conjugated bonds, the spin of an electron interacts with a very large number of nuclei. As a result of this effect, only a single line with a diffuse hyperfine structure is observed in the spectrum. Thus, the spectra are characterized by their integral intensities, depth, shape, and g-factor. Since in all cases the g-factor has a value close to that of a free electron, the other three characteristics of the spectra are usually important.

MAIN CHARACTERISTICS OF THE EPR SPECTRA

The main characteristics of the EPR signals of the majority of semiconducting polymers have been given by Blyumenfel'd et al. [7].

1. Narrow EPR lines are observed in polymers with conjugated bonds.

2. The signals consist of single symmetrical lines, while these lines may have a Lorentzian or a Gaussian or a mixed form. The Gaussian form of the line is observed in systems in which there is no exchange interaction between the isolated regions of the delocalized unpaired spins [7]. The broadening of the line is determined by the interaction of the unpaired spins with protons or with molecules of a paramagnetic impurity, such as oxygen [8]. When exchange occurs, the line acquires a Lorentzian form in the center and a Gaussian form at the edges, and the point of transition

is determined by the exchange frequency. With pronounced exchange interaction, the line has the pure Lorentzian form. An investigation of the form of the line as a function of the conditions of preparation, the temperature of the measurement, and the impurities introduced gives information on the structure and electronic interactions in polymers with conjugated bonds [8-11].

3. The width of the line is between 0.5 and 30 Oe and the g-factor is close to the g-factor of a free electron.

4. Usually, the intensity of the signal rises with an increase in the dimensions of the system of conjugated bonds, although in some cases this is not obligatory. The intensity of the signal may correspond to a concentration of 10^{15}–10^{20} spins/g or one spin per 10^5–10 polymer molecules [7, 12].

5. In the majority of polymers, the dependence of the intensity of the EPR signal on the temperature obeys Curie's law, i.e., the concentration of unpaired spins does not change with a change in the temperature, but, as will be shown below, in a number of cases thermal excitation of paramagnetism is observed.

6. If the polymers contain a linear chain of conjugation and are soluble, the signal is preserved on dissolution and its intensity (calculated per gram) does not depend on the degree of dissolution. If it is assumed that the paramagnetic centers in these polymers are connected with the transfer of charge between neighboring molecules present at distances short in comparison with the neutral crystal, it is possible to explain this effect on the assumption that on dissolution pairs of molecules between which charge transfer takes place are preserved. The concentration of these complexes is determined by the following expression [6]:

$$\frac{n^2}{N_0 - n} = N_0 e^{\frac{-\varepsilon_i^s}{kT}},$$

where n is the concentration of the complexes, N_0 is the concentration of the pairs in solution, ε_i^s is the difference in the energies of the neutral and polar states, and $\varepsilon_i^s = I - A - W(R_0^s) - W_s$, where I is the ionization potential, A is the electron affinity of the neutral molecule, $W(R_0^i)$ is the energy of Coulomb interaction in the polar state, and W_s is the energy of solvation of an ion pair.

7. Various additives, including adsorbed gases, have a considerable influence on the characteristics of the EPR signal. The influence of oxygen on the semiconducting properties and the EPR spectra of polymers with conjugated bonds has been studied in particularly great detail.

In spite of the generality of the observed electronic and paramagnetic characteristics for all polymers with conjugated bonds, there are weighty reasons for separating polymers with semiconducting properties into two groups in each of which completely different mechanisms of excitation and the transfer of the charge carriers act.

In the first group may be placed polymers to which the ideas developed for molecular crystals are applicable. These include polymers with a linear system of conjugated bonds. It can be shown [13] that the growth of the polymeric chain is not accompanied by an increase in the size of the conjugated system and the segments of conjugation have approximately the same structure and length in different samples regardless of their molecular weights. Under these conditions, in spite of the splitting of the electron levels of the individual structural units composing the polymeric molecule, they do not form a quasicontinuous band covering the whole molecule; and the distances between the neighboring levels remain large in view of the limited length of the segments of continuous conjugation. Dispersion forces act between the molecules and if bands covering the polymeric molecules are in fact formed, they are very narrow and their width does not exceed kT (cf. Chapter IV). Generally, these polymers have a conductivity close to that of insulators. This group includes polymers forming donor–acceptor complexes with molecules of added acceptors or forming intramolecular complexes.

The formation of donor–acceptor polymer complexes takes place in two ways. In the first case, a system of conjugated bonds forms a polymer chain which possesses donor properties [14-17], and the complex is formed between the conjugated structural units of the main chain (the number of which usually does not exceed 10) and low-molecular-weight acceptor molecules. In the second case [18-22], the polymer matrix is characterized by a carbon chain with saturated bonds with side-chains (benzene, naphthalene, anthracene, and pyridine radicals, acetate groups, etc.) exhibiting donor properties and forming charge-transfer complexes with the acceptor molecules introduced. This group can be discussed from the point of view of the ideas developed [12, 23-26] on the relative distribution of the levels of polar and nonpolar excitations and their influence on the connection between the concentrations of unpaired electrons and free carriers which will be given below.

To the other group belong polymers with a developed system of conjugated bonds linked into a single molecular network. In one of the stages of their production a considerable temperature effect is necessary. These polymers are insoluble. Their conductivity is that characteristic of semiconductors or semimetals. The energy structure of the electron levels is completely different. An individual region of polyconjugation contains 10^5 or more carbon atoms. On the splitting of the levels, the distance between them is less than the thermal broadening at ordinary temperatures and therefore the formation of a quasicontinuous band may be assumed [27]. The formation and movement of the carriers in the sections of polyconjugated bonds takes place by the ordinary band mechanism and the total transfer of charge carriers also includes transfer from one region to another. We must note immediately that the applicability of the band scheme in this case requires rigorous experimental and theoretical substantiation but it can be stated that in highly conducting polymers no characteristics have been found that would contradict the band model or which could not be explained by it. The division of the polymers into two sharply delimited groups is not only connected with the convenience of considering their electrical and paramagnetic characteristics. It can be seen that these two groups are characterized by different forms of compensation effect, as was shown in Chapter I, i.e., they possess different conduction mechanisms.

It goes without saying that in each of the two groups of polymeric semiconductors described, it is possible to find concrete substances which are characterized by one of the relationships between the electrical and paramagnetic properties described above.

ELECTRONIC STATES OF MOLECULAR CRYSTALS

Let us consider the electrical and magnetic characteristics of systems with conjugated bonds forming molecular crystals [13, 23-26]. It is possible to explain the main features observed in polymers with a linear system of conjugated bonds for which these ideas are applicable (first group of polymers).

Let us consider the potential curves for systems of two molecules in the neutral state and in the formation of a charge-transfer complex (Fig. 11) [26]. In view of the equivalence of the dispersion forces and the forces of repulsion for the neutral and ionic states the potential curves have a point of intersection, R_1:

$$R_0^i < R_1 < R_0^0.$$

The neutral state is an unexcited singlet state. Since terms of the same multiplicity and symmetry cannot intersect, near the point of intersection there is a displacement of the ionic and neutral states as a result of which the difference in the energies of these states increases and the equilibrium distance for the lower level shifts towards the point of intersection $R_0 < R_0^0$.

The arrangement of R_0 and the energy levels is determined by the ratio between the interaction energy H_{12} and the energies of the initial ionic and neutral states. At $R_0 > R_1$ the ground state is close to the neutral state and at $R_0 < R_1$ it becomes ionic. For the excited states, the reverse situation is valid. These ideas can be extended to the case of a molecular crystal. The curve of the neutral state will describe the change in the energy of the dispersion

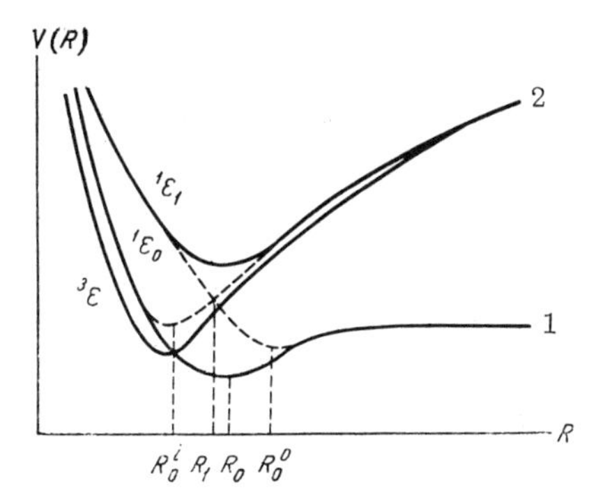

Fig. 11. Potential curves of the neutral (1) and ionic (2) states of two molecules.

interaction of two neighboring molecules in the crystal, which is correct because of the additivity of the dispersion forces. The points of the curves of the ionic states will now correspond to the transfer of an electron between neighboring molecules of the crystal. The state then arising is a quasistationary state, since the transfer exciton formed can migrate through the crystal and the electron must be distributed over all the molecules of the first coordination sphere. Thus, thanks to the strong bond of the valence electrons with the molecules the relationships given above will also be valid for a molecular crystal and will show a change of the energy of the ionic and neutral states of a binary pair of neighboring molecules as a function of the lattice period. Under these circumstances, the energy of the Coulomb interaction $K_{12}^{(1)}$ in the expression for the energy of the ion pair must be replaced by the total energy of the charges in the crystal:

$$W_{12} = \frac{\alpha}{K_0} K_{12}^{(1)} + \left(1 - \frac{\beta}{K_0}\right) E_0,$$

where α and β are the lattice coefficients for the Coulomb energy and the polarization energy, K_0 is the effective dielectric constant of the crystal, and E_0 is the energy of polarization of neighboring molecular ions.

The location of the conduction band relative to ε_1 (the width of the spectrum of the polar state) is determined by the interaction potential of neighboring ions, which depends mainly on the packing density of the crystals and the dimensions of the molecules. With a decrease in R_0, the widths of the spectrum of the polar states decrease in view of the increase in the energy of polarization and H_{12}. Each of the levels is split into a band the width of which is proportional to the energy of interaction H_{12} and increases with a decrease in R_0. Starting from the relative location of R_0, R_0^0, R_0^i, and R_1 it is possible to distinguish the following four types of organic semiconductors forming molecular crystals (Fig. 12) [26]:

1. Weak interaction:

$$R_0 \simeq R_0^0;$$
$${}^1\varepsilon_1 - {}^1\varepsilon_0 = \varepsilon_1 \gg H_{12}.$$

The excited ionic state is fairly high and its interaction with the ground neutral state can be neglected. The triplet level is

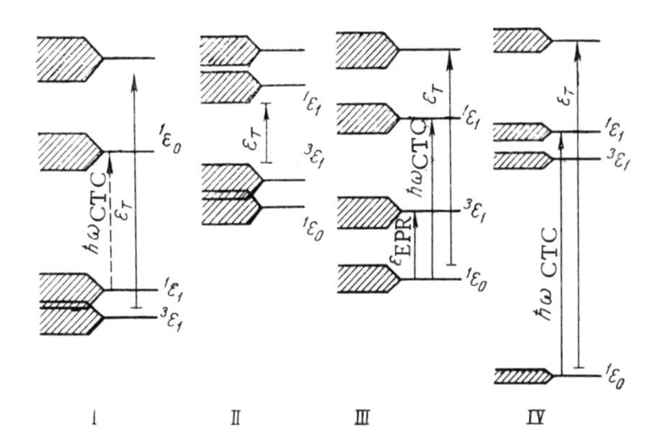

Fig. 12. Energy spectra of organic semiconductors. I) State in an ionic crystal; II) strong interaction, ground state ionic; III) strong interaction, ground state neutral; IV) weak interaction.

located close to the charge-transfer band and, consequently, substances of this type must be diamagnetic. The observed paramagnetism can be connected only with impurity centers. The spectrum of the polar states is located above $^1\varepsilon_1$ and transitions to polar states are unlikely. The observed conductivity is intrinsic and is not connected with paramagnetic centers.

2. Strong interaction, neutral state:

$$R_0 \geqslant R_1, \quad {}^1\varepsilon_1 - {}^1\varepsilon_0 = \varepsilon_1 \simeq H_{12}.$$

The triplet level is closer to the ground state than to the excited state and thermal excitation of paramagnetism is observed. Since the distance between the levels in the spectrum of the polar states becomes comparable with H_{12}, transitions to higher polar states are permitted, which leads to a broadening of the charge-transfer band. The activation energy of conduction is always greater than the energy of thermal excitation to the triplet state (EPR signal). The characteristic sign of these crystals is decreased intermolecular distances. In addition to the thermally excited paramagnetism and the intrinsic conductivity of the main substance, the EPR signals of impurity paramagnetic centers obeying the Curie law and impurity conduction due to their thermal ionization may be observed.

3. Strong interaction, ground state ionic:

$$R_0 \leqslant R_1, \quad \varepsilon_1 \simeq H_{12}.$$

The lowest triplet and singlet states are interchanged and the activation energies of conduction are minimal. With a sufficiently large intermolecular interaction the overlapping of the bands is possible and the activation energy of conduction may become zero. In this case paramagnetism may be ascribed to a free-electron gas and does not depend on the temperature.

4. Ionic compounds:

$$R_0 \simeq R_0^i, \quad \varepsilon_1 \gg H_{12}.$$

The ground state is paramagnetic and obeys the Curie law. The activation energies of conduction again increase.

CONNECTION BETWEEN ELECTRIC AND PARAMAGNETIC PROPERTIES OF POLYMERIC SEMICONDUCTORS TO WHICH THE IDEAS DEVELOPED FOR MOLECULAR CRYSTALS ARE APPLICABLE

Let us now consider the data available on the electrical and paramagnetic characteristics of polymeric semiconductors with linear systems of conjugated bonds with a conductivity close to that of insulators [28-30] and belonging to type 1.

A parallel study of the electrical and paramagnetic characteristics of the polyphenylacetylenes [12] has shown that the intensity of the EPR signal, which is easily detected in these polymers, obeys the Curie law. The central part of the line is Lorentzian and its outer part describes a Gaussian curve. An analysis of the shape of the line according to the theory of exchange narrowing has enabled the number of equivalent protons n over which an unpaired electron is delocalized to be estimated. It was found that

$$n \leqslant 2 \ln 2 \left(\frac{27.2}{\Delta H_{12}} \right)^2 = 11.$$

The value of the equivalent protons obtained permits the statement that the paramagnetic centers are connected with local defects in the structure and the paramagnetism is an impurity (extrinsic) phenomenon. Now let us pass to an account of the results of an experimental investigation of conductivity in connection with paramagnetism. As is well known, the conductivity of a polymer may be intrinsic or extrinsic (due to impurities). Because of the high activation energy of conduction, 2.15 eV, it is difficult to make a choice between the intrinsic and the extrinsic or impurity mechanisms for the generation of free charge carriers. However, this can be done by investigating the optical characteristics of these polymers. The absence of the intrinsic mechanism of generation is indicated by a noncorrespondence of the activation energy of conduction with the long-wave fall in the optical absorption spectrum. The study of the luminescence spectra of the polyphenylacetylenes [13] has enabled the presence of impurity luminescence with a maximum in the 2.0–2.3 eV region to be recognized, which agrees well with the activation energy of dark conduction.

In this system, the effective energy of transfer to luminescence centers is observed. The concentrations of paramagnetic centers and impurity luminescence centers agree to an order of magnitude ($\sim 10^{17}$ g^{-1}). The results obtained enable the formation of free charge carriers in these polymers to be connected with the ionization of impurity paramagnetic centers and a direct correspondence between the electrical and paramagnetic characteristics to be established. Further studies are necessary to determine the concrete structure of the paramagnetic centers.

Let us consider polymers forming weak donor–acceptor complexes with acceptor molecules introduced into the polymer [18–20], which are described by Mulliken's theory. Thus, the complex of polyvinylcarbazole with tetracyanoquinodimethane has a high specific resistance at room temperature ($\sim 10^{15}\,\Omega\cdot$cm) and a high conduction activation energy ($\sim$ 1.3 eV) [20]. In these complexes a high concentration of paramagnetic centers which does not depend on the temperature is observed. When the concentration of acceptor in the polymer is increased, there is a rise in the electrical conductivity and a decrease in the activation energy with a simultaneous increase in the concentration of paramagnetic centers [17].

A detailed consideration of the electrical and paramagnetic properties has been made on polyazines and polymeric Schiff bases forming molecular donor–acceptor complexes with bromine and iodine [15, 16]. In addition to the usual paramagnetic and electrical characteristics (observance of the Curie law for the intensity of the EPR signal, mixed form of the line, exponential dependence of the conductivity on the reciprocal of the absolute temperature, etc.), it has been possible to establish a correlation between the paramagnetic and electrical properties of the substances studied. In the case of amorphous polymers, an increase in the content of acceptor leads to a rise in conductivity and in the concentration of paramagnetic centers and to a fall in the activation energy. The dependence of the conductivity and the activation energy on the content of acceptor has an extreme nature, the finding of a constant concentration of paramagnetic centers corresponding to the content of acceptor at which the maximum conductivity is observed.

In the formation of the complexes, complete charge transfer, recorded in the EPR spectra, takes place only in a few defective sections of the polymer chain where, in view of disturbances of the structure, the acceptor and donor molecules are located at shortened distances. The superfluous donor and acceptor molecules form weak complexes not recorded by EPR. This is confirmed by the fact that, on pumping out, the molecules of halogen-forming non-charge-transfer complexes are desorbed first and the components of the strong complexes subsequently, as can be followed from the EPR spectra. The fact that weak complexes are in fact formed is shown by the appearance in the optical spectrum of a band corresponding to charge transfer and by the appearance of additional intensity of the EPR signal when a polymer containing halogen is illuminated with light of the wavelength corresponding to the absorption maximum. The electrical and paramagnetic properties characterizing the complexes make it possible to postulate an impurity mechanism for the generation of the carriers and extrinsic paramagnetism. No intrinsic conductivity is detected in the measurements and, apparently, these complexes may also be assigned to type 1.

A comparison of the electrical and paramagnetic properties for a large number of polymeric complexes in which the donor components are polymeric substances characterized by the presence of phenyl, pyridine, pyrrolidone, and carbazole substituents and the acceptors are the halogens I_2, Br_2, and IBr, and the halogen compounds of the metals [21, 22] permits these complexes also to be assigned to type 1. However, it is also possible to obtain additional information. On the basis of electrical measurements the polymers studied can be arranged in a sequence in the order of increasing conductivity on the formation of complexes with one of the acceptors. The greatest conductivity and intensity of the EPR signals are observed for polymers possessing well-defined donor properties and forming strong charge-transfer complexes. Analogous "activity" series can be constructed for the acceptors. It is found that steric factors have an influence on the strength of the complex formed which is not less than the donor and acceptor characteristics of the components. For the strong complexes there is an increase in the conductivity of approximately fifteen orders of magnitude as compared with the initial polymer, the absolute conductivity at room temperature amounting to $\sim 10^{-3}\,\Omega^{-1}\cdot cm^{-1}$. A confirmation of the formation of strong charge-transfer complexes is the hypsochromic shift of the absorption band of iodine. In the spectra of solutions of complexes of poly-(vinyl alcohol) with iodine, the hypsochromic shift of the absorption band is very large and does not depend on the iodine content (λ_{max} = 401 nm). This shows the strong charge-transfer interaction of poly(vinyl alcohol) and iodine even in solution, the acceptor being almost completely in the ionic form (I_2^-). It may be concluded that in this case the ground state is ionic and these complexes belong to type 3, since on passing from solutions to the solid state the strength of the complexes can only rise [22].

EPR SPECTRA OF IMPURITIES NOT AFFECTING THE ELECTRICAL CONDUCTIVITY IN HIGHLY CONDUCTING POLYMERIC SEMICONDUCTORS

Let us now consider the relatively highly conducting polymers with conjugated bonds to which we can no longer apply the considerations valid for molecular crystals.

Although formally these polymers are also closest to type 3 of the classification considered and it may be assumed that in each sufficiently extended region of polyconjugated bonds the conditions assumed for this type are satisfied, nevertheless for highly conducting polymers the main factor determining the electrical characteristics is the process of the transfer of charge carriers from one region to another, which is not connected with their paramagnetic characteristics. An attempt may be made to establish such a connection in highly conducting polymers separately.

As an example we may give the polymeric semiconductors obtained by the radiation-thermal treatment of polyethylene. A detailed study of the paramagnetic and electrical characteristics during their gradual change with an increase in the temperature of treatment has led to the following main conclusions [11]. The unpaired spins recorded in the EPR spectra cannot be current carriers, since the concentration of the current carriers calculated from the electrical measurements is lower than the concentration of unpaired spins by several orders of magnitude for samples obtained at temperatures no higher than 800°C (Fig. 13). At the same time, the paramagnetic centers cannot belong to acceptor centers for the following reasons.

As will be shown below, for specimens obtained at temperatures higher than 440°C the conduction activation energy at room temperature is determined by the mobility of the charge carriers. The activation energy of generation of the current carriers is close to zero and the concentration of paramagnetic centers should be close to the concentration of carriers in order of magnitude. This, however, is not the case. A detailed study of the paramagnetic characteristics and a comparison of them with the optical and electrical properties makes most probable the conclusion that the EPR signal for specimens obtained at treatment temperatures

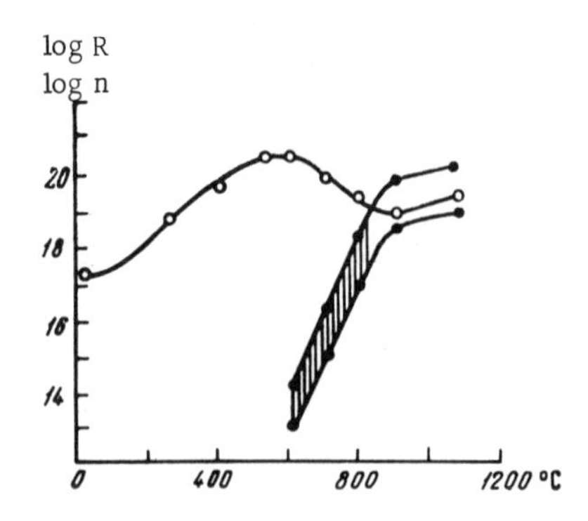

Fig. 13. Concentrations of paramagnetic centers (○) and charged carriers (●) for modified polyethylene as functions of the temperature of treatment.

lower than 800°C is due to structural defects not appearing clearly in the electrical conductivity. Nevertheless, these centers may make their own contribution to the electrical properties [31]. When a sample obtained at a temperature of 620°C is irradiated with fast electrons, a 40-fold decrease in the concentration of paramagnetic centers and an increase in the electrical conductivity by an order of magnitude take place while the activation energy remains constant. However, on heating to 200°C the electrical and paramagnetic characteristics are restored to their initial values. Irradiation with β-particles leads to the formation of electron–hole pairs and, in addition to the recombination process, the process of the capture of electrons and holes by paramagnetic traps takes place. In this process, the concentration of paramagnetic centers decreases and the excess concentration of carriers makes a contribution to the increase in conductivity. On heating, the carriers are expelled from the traps with subsequent recombination, which is accompanied by the restoration of the original values of the concentrations of paramagnetic centers and charge carriers. The necessity for large doses of radiation, at which the filling of the traps take place, is difficult to explain at the present time since we know neither the trapping cross section of the electrons by the paramagnetic traps and the nonparamagnetic recombination centers nor the concentrations of these centers. From this example it can be seen that the paramagnetic centers may be fairly effective traps for charge carriers.

EPR SPECTRA RECORDING IMPURITY STATES IN POLYMERIC SEMICONDUCTORS

Let us now consider a case in which the paramagnetic characteristics are more directly connected with the charge carriers, namely, they characterize in some way or another the centers from which the carriers are formed. The latter statement is fully justified by the fact that the concentration of paramagnetic centers usually exceeds the concentration of free charge carriers evaluated from various measurements by several orders of magnitude. Consequently it is natural that the paramagnetic centers determine the concentration of free carriers if, of course, a connection between them is established experimentally.

In a number of papers [32–36] the connections between charge carriers and unpaired spins have been compared from the point of view of ideas on so-called eka-conjugation. Eka-conjugation refers to such considerable systems of conjugated double bonds that excited electronic states (triplet excitons) exist in appreciable concentration at room temperature. In a comparison of conductivity and paramagnetism for a polyacene quinone, use is made of the theory of absolute reaction rates. The following expression is obtained for the conductivity:

$$\sigma = |e|^2 n_0 \frac{Ld}{3h} \exp\left[\frac{\Delta S_0^{\ddagger} T - \Delta H_0^{\ddagger} - E_0}{kT}\right] \exp\left[\frac{p^{1/2}}{kT}(b''t + b_0)\right],$$

where e is the electronic charge, n_0 is the maximum achievable concentration of carriers, L is the length of molecular chain, d is the width of the barrier through which a carrier tunnels, $S_0^{\ddagger}$ and $H_0^{\ddagger}$ are the entropy and enthalpy of the activated state, E_0 is the energy of formation of pairs of carriers from two neutral molecules, p is the external pressure, and b" and b_0 are deformation constants describing the intermolecular barriers for tunneling.

It is found that between the concentration of unpaired spins N and the conductivity σ the relation

$$N = 7 \cdot 10^{20} \sigma^{1/2}$$

exists; in addition, thermally excited paramagnetism is observed:

$$N = N_0 \exp\left(-\frac{\Delta E_0}{kT}\right),$$

where

$$\Delta E_0 \simeq 0.01 - 0.025 \text{ eV}$$

It is assumed that the charge carriers are formed by a two-stage mechanism the first stage of which is the formation of a triplet exciton:

$$R \rightleftarrows \cdot R \cdot .$$

and

$$\frac{[\cdot R \cdot]}{[R]} = K_1 = \frac{k_{1f}}{k_{1b}} = e^{-{}^3E/kT}.$$

The subsequent formation of pairs of charged carriers takes place as a result of a one- or two-exciton process:

$$R + \cdot R\cdot \rightleftarrows R_0^+ + R_0^-,$$

$$\frac{[R_0^+][R_0^-]}{[R][\cdot R\cdot]} = K_2 = \frac{k_{2f}}{k_{2b}} = e^{-\Delta F_2/kT}, \quad \Delta F_2 = \Delta E_g - {}^3E.$$

$$2\cdot R\cdot \rightleftarrows R_0^+ + R_0^-,$$

$$\frac{[R_0^+][R_0^-]}{[\cdot R\cdot]^2} = K_2' = \frac{k_{2f'}}{k_{2b'}} = e^{-\Delta F_2'/kT}, \quad \Delta F_2' = E_g - 2{}^3E,$$

where ${}^3E$ is the energy of formation of the biradical state; E_g is the energy of formation of a pair of carriers, and $k_{1f}, k_{1b}, k_{2f}, k_{2b}, k_{2f'}$, and $k_{2b'}$ are the specific constants of the reaction rates.

The two processes lead to a single relation at $R_0^+ = R_0^-$:

$$[R_0^-] = [\cdot R\cdot]\left(\frac{K_2}{K_1}\right)^{1/2}.$$

The detailed consideration of the electronic processes in a solid from the point of view of the theory of absolute reaction rates is clearly descriptive, but it permits high concentrations of unpaired spins to be connected with reasonable values of the concentration of carriers (which, however, can also be done, as will be shown, by using the ordinary concepts of the physics of solids).

In actual fact, with the measured values $E_g \simeq 0.4$ eV, ${}^3E \simeq 0.02$ eV, $\sigma_{25} = 6 \times 10^{-5}\,\Omega^{-4}\cdot\text{cm}^{-1}$, and $N = 10^{20}\text{cm}^{-3}$ it is possible from the scheme given to calculate the concentration of carriers n, which proves to be $7 \times 10^{16}\text{ cm}^{-3}$. The figure obtained agrees with the values calculated from the band model.

Similar conclusions have been arrived at by Nechtschein [37]. In a study of the temperature dependences of the paramagnetic and electrical characteristics of pyrolyzed polyacrylonitrile, the appearance of unpaired spins with an excitation energy of not more than 10^{-5} eV, which corresponds to the ground state, was noted. It may be considered that in the ground state there is a mixture of singlet and triplet states, the presence of triplet states explaining the paramagnetism of polymers with conjugated double bonds.

A more far-reaching study of the EPR spectra and the electrical conductivity with the object of relating the high concentration of spins to the concentration of carriers has been carried out by Vlasova et al. [38, 39] on highly conducting polymers obtained by chemical synthesis and by transformations in the chains of macromolecules. Polypyridines with the structural formula (—N=C—C=C—C=C—) with different degrees of polymerization and polyacrylonitrile thermolyzed at temperatures between 200 and 725°C were studied.

In the investigation of these two polymers, identical laws were established. Let us consider only the polypyridines. It was natural to compare the paramagnetic characteristics with the concentration of free charge carriers, information on which was obtained from measurements of the thermo-emf. All the specimens were characterized by a single symmetrical line of Lorentzian form with $\Delta H = 2$–7 Oe. The intensity of the absorption lines obeyed the Curie law, which shows that the concentration of paramagnetic centers is independent of the temperature of measurement. In all cases the thermo-emf was negative and had different temperature dependences. The concentration of carriers was calculated on the basis of the assumptions that the band model is correct, the scattering of the carriers takes place on the thermal vibrations of the lattice, and the effective mass m^* is equal to the mass of a free electron m_0. From the usual formulas of the band theory it is possible to find the activation energy for the generation of carriers ΔE_n and their concentration n for the various specimens given in Table 5.

It is, further, possible to assume that the paramagnetic centers are formed by a donor impurity, i.e., $N = N_d$, and to calculate the thermo-emf with this condition. For samples in which at temperatures in the measured range

$$e^{\Delta E_n/kT} \simeq 2\,(2\pi mkT)^{3/2}/N_d \cdot h^3,$$

the following general formula is valid:

$$\alpha = \frac{k}{e}\left\{2 + \frac{\Delta E_n}{kT} - \ln\frac{1}{2}\left[\sqrt{1 + 2e^{\Delta E_n/kT}\cdot N_d\, h^3/(2\pi mkT)^{3/2}} - 1\right]\right\}.$$

When

$$e^{\Delta E_n/kT} \gg 2\,(2\pi mkT)^{3/2}/N_d \cdot h^3,$$

TABLE 5. Concentration of Free Carriers and Number of Paramagnetic Centers (N) in Polypyridines Obtained under Various Polymerization Conditions

Temperature, °C	Time, h	N, cm^{-3}	ΔH, Oe	Thermo-emf (in μV/deg at 21°C	ΔE_n, eV	n, cm^{-3}
330	5	$1.6 \cdot 10^{19}$	6.4	−505	0.15	$9.8 \cdot 10^{17}$
330*	10	$1.5 \cdot 10^{19}$	6.0	−520	0.2	$3.5 \cdot 10^{17}$
330	10	$1.4 \cdot 10^{19}$	6.5	−650	0.3	$4.5 \cdot 10^{16}$
330	25	$5.4 \cdot 10^{19}$	4	−195	0.05	$1.2 \cdot 10^{19}$
340	8	$3.5 \cdot 10^{19}$	5	−300	0.09	$5.4 \cdot 10^{18}$
350	24	$7.6 \cdot 10^{19}$	2.5	− 95	0	$7.6 \cdot 10^{19}$
370	10	$1.1 \cdot 10^{20}$	2	−100	0	$1.1 \cdot 10^{20}$
Polymerization of pyridine hydrochloride		$9.3 \cdot 10^{19}$	2	−95	0	$9.3 \cdot 10^{19}$

* Specimen treated with atmospheric oxygen.

calculation is carried out by the usual simplified formula

$$\alpha = \frac{k}{e}\left\{2 + \ln\frac{2^{1/2}(2\pi mkT)^{3/4}}{h^{3/2}} - \ln N_d^{1/2} + \frac{\Delta E_n}{2kT}\right\}.$$

When $n = N_d$ = const, a formula is used which takes into account the degeneracy of the charge carriers:

$$\alpha = \frac{k}{e}\left\{\frac{2F_1(\mu^*)}{F_0(\mu^*)} - \mu^*\right\},$$

where $F(\mu)^*$ are the Fermi integrals and $\mu^* = \mu/kT$ (where μ is the chemical potential of the electrons) since for this case the degeneracy condition $h^2(3\pi^2 n)^{3/2}/2\ mkT \gg 1$ is satisfied.

The numerical values of α obtained agree well with those measured both in absolute magnitude and in their temperature dependence. The assumption that unpaired electrons play the role of a donor impurity is also illustrated by Fig. 14, which gives, together with the theoretical dependence of α on n, the experimental dependence of the thermo-emf on the concentration of charge carriers obtained from known values of $n = N_d$ and ΔE_n. As can be seen, the experimental points lie well on the theoretical curve. For thermolyzed polyacrylonitrile, the same correspondence is obtained at $m^* = 1.2\ m_0$. In such considerations it is tacitly as-

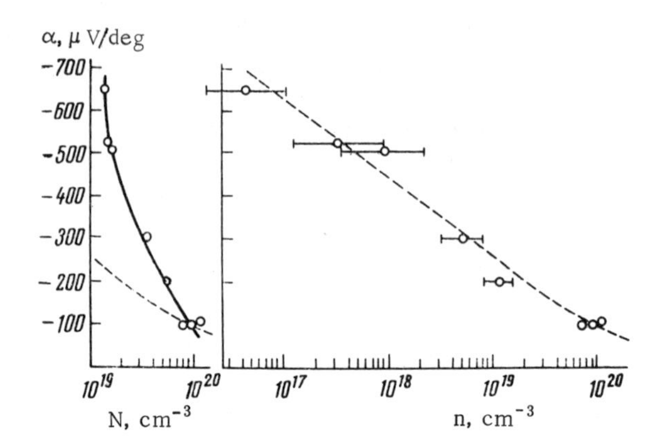

Fig. 14. Dependence of the thermo-emf on the concentration of unpaired electrons N and the concentration of charge carriers n. Broken lines — theoretical dependence of α on the concentration of charge carriers.

sumed that the free electrons arising in the conduction band on the ionization of the donors give precisely the same EPR signal as the un-ionized donors.

In actual fact, it is difficult to imagine that after the ionization of the donor, i.e., the departure of one electron from the outer electron cloud, the paramagnetic characteristics of the donor center, which are determined by the same outer cloud, remain unchanged. Consequently, we should observe, in the first place, a substantial difference in the EPR spectra of samples with completely ionized donors, $N_d = n(\Delta E_n = 0)$, and almost un-ionized donors, $N_d \gg n(\Delta E_n \gg 0.15)$ and, in the second place, thermally excited paramagnetism for samples with $N_d \gtrless n$. Since the EPR spectra of various specimens have similar characteristics and in all cases the intensity of absorption obeys the Curie law, it follows that the signals from the un-ionized donors and the free electrons must coincide. This would be unlikely only in the single case where the donor centers are structural defects in which states with full charge transfer are realized. If, for any reason, the EPR spectra record only the positively charged components of the complexes, the presence of electrons in the bound state in the complex or in the free state — in the conduction band — is not reflected in the EPR spectra. Since the presence or absence of charge-transfer complexes is not shown, additional experiments are required to identify paramagnetic centers with free charge carriers in these materials.

EPR SPECTRA RECORDING CURRENT CARRIERS

Investigations [11, 40] with samples of polymeric semiconductors obtained by the radiation-thermal modification of linear polymers in which the EPR signal is clearly not connected with charge-transfer states have been devoted particularly to the study of the EPR signals caused by the current carriers. Above all, the connection of the paramagnetic and electrical characteristics in materials obtained at treatment temperatures of less than 800°C was considered. On the basis of electrical measurements according to the model that will be considered in subsequent sections, estimates were made of the effective mobility μ and the concentration of positively charged current carriers p as functions of the temperature of treatment of the sample. Since the values of the effective mass of the holes in these substances are unknown but they are, apparently, not less than the mass of a free electron, the values of μ and p can be obtained for the two extreme cases $m^* = m_0$ and $m^* = 10m_0$. Figure 13 shows the concentration of paramagnetic centers in pumped-out samples and the concentration of current carriers as functions of the temperature of thermal treatment. The concentration of current carriers was calculated on the assumption that m^* corresponds to m_0 and 10 m_0 (lower and upper limits of the hatched region). As follows from the figure, at treatment temperatures >800°C the absolute magnitude of the concentration of paramagnetic centers falls into the region of the concentrations of charge carriers, and their changes with a variation in the temperature of treatment coincide. On this basis it was concluded that in the given case it is apparently just the current carriers that are responsible for the EPR signal. From the characteristics of the EPR spectra — the line widths, 3.5 Oe, and the g-factors, 2.0027 — it has been possible to evaluate the relaxation time of the carriers τ_r from the formula for the time of spin-lattice relaxation T_1:

$$T_1 = \left(\frac{3\times 10^{-8}}{\Delta H}\right) = \frac{1}{30}\,\frac{\tau_r}{(g - g_{free})^2}$$

TABLE 6. Mobility and Concentration of Carriers in Radiation-Thermally Modified Polyethylene Measured by Different Methods

Method of measurement	m^*	Temperature of treating the samples, °C			
		930		1100	
		μ, cm²/V·sec	p, cm⁻³	μ, cm²/V·sec	p, cm⁻³
Electrical	$10\ m_0$	17	$2.4\cdot10^{19}$	17	$3.6\cdot10^{19}$
	m_0	115	$3.5\cdot10^{18}$	115	$5.2\cdot10^{18}$
EPR	$10\ m_0$	13	$7\cdot10^{18}$	13	$1.5\cdot10^{19}$
	m_0	130		130	

and further, for the simplest case, to obtain the microscopic mobility:

$$\mu = \frac{e}{m^*}\tau_r$$

Thus, there was the possibility of estimating the mobility and concentration of the carriers by different methods, and their values are given in Table 6.

As can be seen from the table, there is good agreement between the values found. The measured conductivity of the sample obtained at a treatment temperature of 930°C is $56\ \Omega^{-1}\cdot\text{cm}^{-1}$. If the density of the substance is taken as 1 g/cm³, then at the concentration of current carriers determined by the EPR method and equal to $7\times10^{18}\text{g}^{-1}$ the mobility is 50 cm²/V · sec, which corresponds to $m^* = 2.5\ m_0$.

A consideration of the electrical and paramagnetic characteristics of a large number of semiconducting polymers [41–44] leads to similar results. In the production of polyarylenequinones, products of the thermal treatment of poly(methyl vinyl ketone), and polyarylenepolyacetylenes, charge-transfer complexes are formed as a result of structural defects. All the unpaired electrons appearing in the treatment of the complexes are localized on defects and participate in the electrical conductivity by a jump mechanism. Thus, in this case the EPR records the current carriers.

It may be mentioned that there is a large number of papers on the study and comparison of the electrical and magnetic properties of carbonaceous substances obtained at temperatures of more than 1000°C. Such materials do not belong directly to the organic semiconductors.

The results of the comparison of the electrophysical and paramagnetic properties of the polymeric semiconductors that have been given make it possible to draw the following conclusions. Most frequently it is impossible to establish a direct connection between the electronic states of the system which determine the electrical and paramagnetic properties. Sometimes it is impossible on the basis of measurements of the EPR spectra alone to assign the spectra to any particular state participating in conduction. If such a connection does exist, as has been shown for some polymeric semiconductors, its detection requires a detailed study and comparison of the electrical and paramagnetic properties over a wide range of experimental conditions, chemical compositions, and structures of the materials investigated. The necessity for discussing the electrical characteristics in order to understand the connection between them and the observed paramagnetic centers is confirmed particularly clearly by the fact that such a connection has been established more or less reliably for highly conducting polymeric semiconductors, with which the most accurate and diverse electrical measurements are possible.

REFERENCES

1. F. Winslow, W. Baker, and W. Jager, J. Am. Chem. Soc., 77:4751 (1955).
2. B. I. Liogon'kii, A. S. Lyubcheno, A. A. Berlin, L. A. Blyumenfel'd, and V. P. Parini, Vysokomolek. Soedin., 2:1494 (1960).
3. D. Ingram, Free Radicals as Studied by Electron Spin Resonance, Butterworth, London (1958).
4. A. A. Berlin, Khim. Prom., 1962(12):23.
5. E. L. Frankevich and V. L. Tal'roze, Proceedings of the Second All-Union Conference on Radiation Chemistry [in Russian], Izd. Akad. Nauk SSSR, Moscow (1962), p. 651.
6. V. A. Benderskii and L. A. Blyumenfel'd, Dokl. Akad. Nauk SSSR, 144:813 (1962).
7. L. A. Blymenfel'd, V. V. Voevodskii, and A. G. Semenov, Electron Paramagnetic Resonance in Chemistry [in Russian], Izd. Sibirsk. Otdel. Akad. Nauk SSSR, Novosibirsk (1962).
8. A. D. Grishina and N. A. Bakh, Vysokomolek. Soedin., 7:1698 (1965).
9. A. D. Grishina and N. A. Bakh, Zh. Strukt. Khim., 6:198 (1965).
10. A. D. Grishina and N. A. Bakh, ibid., p. 204.
11. N. A. Bakh, A. V. Vannikov, A. D. Grishina, and S. V. Nizhnii, Usp. Khim., 34:1733 (1965).
12. V. A. Benderskii, B. Ya. Kogan, V. F. Gachkovskii, and I. A. Shlyapnikova, in: Carbochain High-Molecular-Weight Compounds, G. S. Kolesnikov (ed.), Izd. Akad. Nauk SSSR, Moscow (1963), p. 253.

13. V. A. Benderskii and P. A. Stunzhas, Vysokomolek. Soedin., 6:1104 (1964).
14. Yu. A. Popov, Candidate's Thesis [in Russian], Inst. Neftekhim. Sintesa Akad. Nauk SSSR, Moscow (1963).
15. G. P. Karpacheva, Izv. Akad. Nauk SSSR, Ser. Khim., 1965:190.
16. G. P. Karpacheva, Zh. Fiz. Khim., 39:3015 (1965).
17. K. Kuwata, Y. Sato, and K. Hiroto, Bull. Chem. Soc. Japan, 37:1391 (1964).
18. W. Slough, Trans. Faraday Soc., 58:2360 (1962).
19. S. Mainthina, P. Kronick, and M. Labes, J. Chem. Phys., 41:2206 (1964).
20. A. Taniguchi, S. Kanda, T. Nogaito, S. Kasabayshi, H. Mikawa, and K. Ito, Bull. Chem. Soc. Japan, 37:1386 (1964).
21. Yu. I. Vasilenok, B. É. Davydov, B. A. Krentsel', and B. I. Sazhin, Vysokomolek. Soedin., 7:626 (1965).
22. Yu. I. Vasilenok, Candidate's Thesis [in Russian], Inst. Neftekhim. Sintesa Akad. Nauk SSSR, Moscow (1967).
23. L. A. Blyumenfel'd and V. A. Benderskii, Zh. Strukt. Khim., 4:405 (1963).
24. V. A. Benderskii, ibid., p. 415.
25. V. A. Benderskii, L. A. Blyumenfel'd, and D. A. Popov, Zh. Strukt. Khim., 7:370 (1966).
26. L. A. Blyumenfel'd, V. A. Benderskii, and P. A. Stunzhas, ibid., p. 686.
27. L. S. Stil'bans and L. D. Rozenshtein, in: Electrical Conductivity of Organic Semiconductors [Russian translation], IL, Moscow (1963), p. 5.
28. A. A. Berlin, L. A. Blyumenfel'd, M. I. Cherkashin, A. É. Kalmanson, and O. G. Sel'skaya, Vysokomolek. Soedin., 1:1361 (1959).
29. L. A. Blyumenfel'd, A. A. Berlin, A. A. Slinkin, and A. É. Kalmanson, Zh. Strukt. Khim., 1:1031 (1960).
30. L. A. Blyumenfel'd, A. A. Berlin, N. G. Matveeva, and A. É. Kalmanson, Vysokomolek. Soedin., 1:1647 (1959).
31. A. D. Grishina, A. V. Vannikov, S. G. Ashin, and N. A. Bakh, Fiz. Tverd. Tela, 9:1651 (1967).
32. H. A. Pohl and L. A. Opp, J. Phys. Chem., 66:2121 (1962).
33. H. A. Pohl and E. Engelhardt, J. Phys. Chem., 66:2085 (1962).
34. H. A. Pohl, A. Rembaum, and A. Henry, J. Am. Chem. Soc., 84:2699 (1962).
35. H. A. Pohl, C. Cappins, and G. Gogos, J. Polymer Sci., A1:2207 (1963).
36. H. A. Pohl and R. Charloff, J. Polymer Sci., A2:2787 (1964).
37. M. Nechtschein, J. Polymer Sci., C2:1367 (1964).
38. R. M. Vlasova and A. V. Airapetyants, Fiz. Tverd. Tela, 7:3079 (1965).
39. R. M. Vlasova, S. N. Gasparyan, V. A. Kargin, L. D. Rozenshtein, and V. E. Kholmogorov, Dokl. Akad. Nauk SSSR, 171:132 (1966).
40. A. D. Grishina and A. V. Vannikov, Dokl. Akad. Nauk SSSR, 156:647 (1964).
41. A. A. Dulov, A. A. Slinkin, A. M. Rubinshtein, and I. L. Kotlyarevskii, Dokl. Akad. Nauk SSSR, 143:1355 (1962).
42. A. A. Dulov, A. A. Slinkin, A. M. Rubinshtein, and I. L. Kotlyarevskii, Izv. Akad. Nauk SSSR, Ser. Khim., 1963:1910.
43. A. A. Dulov, A. A. Slinkin, and A. M. Rubinshtein, Izv. Akad. Nauk SSSR, Ser. Khim., 1964:26.
44. A. A. Dulov, B. I. Liogon'kii, A. V. Ragimov, A. A. Slinkin, and A. A. Berlin, Izv. Akad. Nauk SSSR, Ser. Khim., 1967:909.

CHAPTER IV

MECHANISM OF CONDUCTION IN ORGANIC SEMICONDUCTORS

A large number of investigations has been devoted to the mechanism of conduction in organic semiconductors. The first step in the study of the mechanism of conduction is generally the determination of n and μ appearing in the expression for conductivity $\sigma = en\mu$. In inorganic semiconductors, this is done by measuring the Hall effect. Then, by studying the dependence of μ and n on the temperature and other parameters, the mechanism of conduction in the material investigated is deduced.

In the study of organic semiconductors it was found that in the majority of materials it is impossible to measure the Hall effect and the necessity arose for using other, less traditional, methods. Attempts to obtain information on the mechanism of conduction were made by the measurement of the thermo-emf, frequency dependences, EPR, and the direct measurement of mobility. At the very beginning it must be noted that the first two of these can give only indirect information on the material studied and the final result depends strongly on the model selected.

In spite of many unsuccessful attempts to measure the Hall effect it has nevertheless been possible to observe the appearance of a Hall emf in many materials, the relationships obtained being explained by the band theory.

MEASUREMENT OF THE HALL EFFECT

Within the framework of the usual ideas for semiconductors with a single type of carrier, the Hall coefficient is

$$R = -\frac{3\pi}{8}\frac{1}{en},$$

where n is the concentration of carriers and e is the electronic charge. By measuring the conductivity it is possible from the relation $R \cdot \sigma = \mu_H$ to determine the Hall mobility. Later the mobilities determined from various measurements and having the following sense will be important for us [1].

1. The microscopic mobility, i.e., the rate referred to unit field strength

$$\mu = \frac{v}{E}.$$

2. The drift mobility $\mu_{dr} = d/(T_t \cdot E)$, where d is the distance travelled and T_t is the time of transfer of the injected carriers through distance d.

3. The Hall mobility. For an isotropic non-degenerate semiconductor, $\mu_H = {}^3\!/_8\, \pi\mu$.

4. The ohmic mobility, calculated from the equation

$$\mu = \frac{\sigma}{ne}$$

with the substitution of the concentration of charge carriers.

For semiconductors with a complex structure, these mobility figures may differ markedly. Thus, for polycrystalline samples, transfer from one crystallite to another must be taken into account in the Hall mobility, while the microscopic mobility relates to the motion of the current carriers within the limits of a single crystallite, on the one hand, and the drift mobility is determined by the concentration and energy levels of the carrier traps. In each concrete case it must be stated which mobility is being considered and its relationship with the measured dark conductivity must be taken into account.

Attempts to study the mechanism of conduction in detail have been made on a polyphthalocyanine [2]. However, the results of this investigation have not been confirmed in a study of polyphthalocyanines by other authors [3-5]. It was impossible to measure the Hall effect in these samples, but the mobility and concentration of the carriers were estimated on the basis of measurements of conductivity, taking the influence of oxygen into account, as 10^{-2} cm^2/V · sec and 3×10^{18} cm^{-3}, respectively. It has been possible to measure the Hall effect for a polymeric complex of polyvinylcarbazole with iodine ($\mu_H \sim 0.4$ cm^2/V · sec) [6] and for a number of polyacene quinones [7]. The most reliable results were obtained for a polymer obtained from pyrene and pyromellitic dianhydride, in which a hole mechanism of conduction was found. At room temperature, the Hall constant was 288 cm^3/coulomb, the mobility 0.4 cm^2/V · sec, and the concentration of carriers 2×10^{16} cm^{-3}. The mobility of the carriers calculated from measurements of the Hall effect increased if the material was prepared at a higher temperature.

In measurements of the thermo-emf and the Hall effect of a pyrolyzed ion-exchange resin [poly(acrylic acid) in structural association with 6% of divinylbenzene] with additions of Ca, Na, Th, it was shown [8] that these materials are n-type semiconductors with a concentration of carriers of $1\text{-}3 \times 10^{20}$ cm^{-3} and mobilities of 2-4 cm^2/V · sec. In the case of polymers with inclusions of nickel, the number of carriers remains approximately constant while the mobility increases to 100 cm^2/V · sec.

In spite of the numerous attempts to measure the Hall effect in low-molecular-weight organic semiconductors, it has been possible to do this only for the phthalocyanines. In monocrystals of metalfree phthalocyanine the Hall voltage is at the level of the noise and to isolate it, it was necessary to integrate the voltage as a function of the time on switching on and switching off the magnetic field [9]. Rough estimates of the mobility and concentration of the carriers (n-type) were obtained: 0.1-0.4 cm^3/V · sec and $2\text{-}12 \times 10^6$ cm^{-3}. Higher values of the Hall mobility in crystals of copper phthalocyanine were observed by Delacote and Shott [10]. The Hall effect was measured in the temperature range from 400 to 600°K. According to these measurements, the current carriers are holes.

Clearer results were obtained in a measurement of the Hall effect on monocrystals of copper phthalocyanine by Heilmeier and

TABLE 7. Mobilities for Monocrystals of Copper Phthalocyanine

T, °K	μ_H	μ_n	μ_p
	found	calculated	
300	(−) 75	88	131
322.5	(−) 52	82.3	122
337	(−) 31	77.5	116
357	(−) 16	68	101.5
396	(+) 18	26.2	39

Harrison [11]. In the majority of the samples studied, the sign of the carriers was negative. At 300°K, the Hall mobility was 70 $cm^2/V \cdot sec$. With a rise in the temperature, μ_H fell rapidly to values lying below the sensitivity of the apparatus. For these samples no breaks were found in the curve of the logarithm of the conductivity as a function of the reciprocal temperature. Interesting results were obtained on one of the samples. After the fall in the Hall voltage corresponding to negative charges, with a rise in the temperature an increase in the positive Hall voltage was observed. In measuring the temperature dependence of the conductivity of the same sample it was found that with a rise in the temperature there was a change in the slope of the straight line and ΔE increased from 1.66 to 2.0 eV. This was connected with the presence of adsorbed oxygen in the crystal. If it is assumed that oxygen is a donor impurity, a value of 0.67 is obtained for the ratio of the hole and electron mobilities at 373°K. The dependence of the mobility of the electrons and holes on the temperature can be calculated on the condition that the temperature dependences of the mobilities and the effective densities of the states for holes and electrons coincide. These results are given in Table 7.

With a rise in the temperature, the mobility of the electrons and the holes falls. This shows the applicability of the band model to this system.

From what has been said the conclusion can be drawn that measurements of the Hall effect are very sparse; for organic semiconductors they are generally carried out at the limit of sensitivity of the apparatus and in the best case they are only estimates. The simple expression characterizing the connection between R and n is valid under conditions the fulfillment of which in organic semiconductors requires experimental and theoretical substantiation. Since in low-molecular-weight organic semiconductors, the width of the conduction band is of the order of kT or lower at room temperature, it is impossible to consider that the effective mass does not depend on the energy [12] and in this case the Hall constant cannot be determined from the simple expression given above.

Moreover, if the mobilities of the holes and electrons do not differ greatly, we must take into account the contribution of intrinsic carriers, since in this case the following formula is valid for the Hall coefficient

$$R = -\frac{3\pi}{8}\cdot\frac{1}{e}\cdot\frac{\left(n\mu_n^2 - p\mu_p^2\right)}{(n\mu_n + p\mu_p)^2}.$$

Since the concentration of intrinsic carriers in organic semiconductors is generally unknown, calculations from measured values of R must be treated with some circumspection for this reason, also.

MEASUREMENT OF THE THERMO - EMF

Together with the Hall effect, the thermo-emf is also a classical source of information on the concentration of carriers and the mechanism of conduction in semiconductors. The thermoelectric effects in organic semiconductors have been studied for many years. Some of the first to be studied were polymeric semiconducting products of the thermal treatment of polydivinylbenzene [13] and, among the low-molecular-weight organic semiconductors, phthalocyanines of various metals [14]. In both groups, a weak dependence of the differential thermo-emf determined as the electromotive force per degree of temperature gradient was found to depend slightly on the temperature of measurement. This gave grounds for the first suggestion for low-molecular-weight semiconductors [15] and for polymeric semiconductors [16] that the concentration of carriers in organic semiconductors does not change with a change in the temperature and that the activation dependence of the conductivity on the temperature is determined by the mobility of the charge carriers. Further investigations showed that this conclusion is correct only for relatively highly conducting polymers. For pure crystals of low-molecular-weight organic semiconductors, with the exception of highly conducting charge-transfer complexes, this condition is not normally satisfied. Let us consider the relatively highly conducting polymers, since only in this case can reliable results be obtained in measurements of the thermo-emf.

The most complete measurements have been carried out on thermolyzed polyacrylonitrile [16-19]. For highly conducting samples ($\sigma \geq 10^{-6}\,\Omega^{-1}\cdot\mathrm{cm}^{-1}$), the thermo-emf α obeys the relation

$$\alpha = \frac{k}{e}(A + 2.3 \log T^{3/2}), \quad (1)$$

where T is the temperature, °K.

If it is assumed that the band model is correct, this relation is observed when the concentration of carriers does not change with a change in the temperature in a nondegenerate semiconductor with one type of carrier. For samples with a conductivity less than $10^{-8}\,\Omega^{-1}\cdot\mathrm{cm}^{-1}$ the following equation is valid for α:

$$\alpha = \frac{k}{e}\left(A' + 2.3 \log T^{3/2} + \frac{\Delta E_n}{kT}\right). \quad (2)$$

The last member in equation (2) characterizes the temperature change of the concentration of carriers (activation energy ΔE_n). For all specimens the conductivity depends exponentially on the temperature. For highly conducting specimens, the activation energy of conduction is due wholly to the mobility of the current carriers. For more poorly conducting materials, it is possible to compare the activation energies of conduction ΔE and of the generation of the free carriers ΔE_n. For example, for two samples with conductivities at room temperature of 3.2×10^{-12} and $1.6 \times 10^{-9}\,\Omega^{-1}\cdot\mathrm{cm}^{-1}$, ΔE and ΔE_n are, respectively, 1.4 and 1.0 eV and 0.4 and 0.07 eV [18]. The difference $\Delta E - \Delta E_n$ is the activation energy of the mobility. Thus, in poorly conducting polymeric semiconductors, unlike the highly conducting materials, the concentration of carriers increases with a rise in the temperature, but in this case the activation dependence of the conductivity is determined mainly by the temperature change in mobility. Small values of ΔE_n determine a very small temperature change in α, as is generally observed.

What has been stated above shows the possibility of using the band scheme for explaining the electrophysical characteristics of regions of polyconjugated double bonds. However, in quantitative calculations using measured values of α (for example, of the concentration of carriers), great circumspection is necessary, since the contribution of the intrinsic carriers to the thermo-emf is generally unknown. Moreover, for α to characterize the highly conducting regions one must assume that the dimensions of the regions of polyconjugation considerably exceed the dimensions of the nonstructured layers and that the thermal conductivities in both are approximately the same. The first condition is always satisfied but the second is not necessarily satisfied. It may be assumed that the thermal conductivity of the regions of conjugated bonds is higher than the thermal conductivity of the unstructured intervals because of the electronic component connected with conductivity. In this case, the temperature fall over regions of polyconjugated bonds will be less than the measured temperature gradient in the sample.

A slight increase in α with a rise in the temperature, or its constancy, is also observed in a number of other semiconducting polymers [20-23]. The characteristics of the mobility and of the concentrations of charge carriers obtained on the basis of measurements of the thermo-emf depend greatly on the model selected. Different and, in some cases, identical polymers are considered dissimilarly by different authors: some propose an extrinsic (impurity) mechanism of conduction [17-19] and others an intrinsic mechanism [24-26].

STUDY OF THE ELECTRICAL PROPERTIES IN AN ALTERNATING CURRENT

For low-molecular-weight organic semiconductors, extremely large dielectric losses have been observed which are not compatible with the results of a study of conduction obtained in measurements with direct current [27]. Garrett, in his review (see [28]), was the first to draw attention to this fact. Subsequently, in a number of investigations [29-33] it was found that resistance in an alternating current may be less by several orders of magnitude than in direct current. In a study of polymeric complexes of tetracyanoethylene the hypothesis was put forward that this may be connected with the polarization of the macromolecules. A more detailed investigation of polymeric complexes of tetracyanoethylene has shown that these films possess an extremely high effective dielectric constant which falls with an increase in frequency [31].

In the study of the resistance of a film of a polymeric complex of tetracyanoethylene with silver in the range of frequencies from 0 to 200 MHz it was found that the resistance in the range studied decreases by two orders of magnitude with an increase in the frequency and then, beginning from 10 MHz and continuing up to 200 MHz it is independent of the frequency [22]. The same characteristics have been observed for the frequency dependence of the resistance of radiation-thermally modified polyethylene, [34, 36]. The thermal activation energy depends on the frequency at which the measurements are carried out. With an increase in the frequency the activation energy of conduction of a metalfree film decreases. A similar situation has been observed in a study of the properties of radiation- and heat-treated polyethylene.

A study over a wide frequency range of samples with different specific resistances has shown that the dependence of the resistance on the frequency is greater the higher the specific resistance. With an increase in the electrical conductivity, the slopes of the curves decrease. The behavior of the effective dielectric constant and the specific resistance described above are generally characteristic of materials in which the movement of the current carriers is limited by barriers of any shape and dimension between the highly conducting regions. In order to pass from one region of continuous conjugation to another, an electron must overcome a barrier at the boundary. The nature of the barriers may differ and is generally not accurately known. It may be considered that the barrier is a separate phase, a substance conducting worse than the bulk of the film. It is possible that the barriers appear as a consequence of a contact potential difference between regions of continuous conjugation of different dimensions [22, 31]. Obviously, the materials studied contain both types of disturbances. Under these conditions, at lower frequencies phase barriers arise and at frequencies of 10^6-10^7 Hz the properties characterizing the regions of continuous conjugation appear.

MODEL USED TO STUDY THE MECHANISM OF CONDUCTION IN POLYMERIC SEMICONDUCTORS

The results given can be considered with the aid of three types of model between which there are no sharp boundaries.

In the first place, this consists of point impurities in some continuum, which corresponds to the model used in considering jump conduction by impurities in germanium and silicon. In such a model, the frequency characteristics of the active and reactive components are directly connected with the frequencies of the jumps of the carriers between the impurity centers [36-37]. On the superposition of an alternating current of such a frequency that a carrier can perform relaxation oscillations following the field, an active component of the current with zero activation energy appears, since the relaxation oscillations take place without activation.

Let us now assume that the impurity cannot be regarded as a point impurity and the distance between the impurities is comparable with their dimensions. This is now the heterogeneous model considered for the case of organic semiconductors in a number of papers, for example [29, 34, 38]. In agreement with the concrete structure in this type of model, it is assumed that the relaxation oscillations of the carriers take place within the limits of regions of continuous conjugation, and from the frequency relationships it is possible to obtain the mean distance between such regions and the dimensions of the regions themselves [38].

Finally, in the third type of model it is assumed that the distance between the "impurities" is smaller than the dimensions of the "impurities" themselves. This corresponds to regions of continuous conjugation separated by thin barriers. If it is assumed that the carriers move in the macromolecule as free particles under the action of the alternating field, it can be seen that, beginning at that frequency at which the electrons cannot traverse a distance greater than the length of continuous conjugation in the time of a half-period, the resistance will remain constant. In practice, the polymer consists of a set of macromolecules of different lengths. This conclusions has been drawn on the basis of the wide spectrum of relaxation times (from 10^{-2} to 10^{-7} sec) observed in the study of various materials. In this case, the shorter macromolecules are electron acceptors with respect to the longer ones, as follows from data on the work function of an electron [39, 40].

Thus, we have a substance consisting completely of "impurities" with both donor and acceptor characteristics. The peculiarity of such "impurities" consists in the fact that they themselves possess a large number of carriers, free only within a macromolecule which, however, can be detected in alternating-current measurements.

A well-known electrotechnical analog of the process in which the movement of the current carriers is retarded by the grain boundaries or barrier layers is the Maxwell–Wagner condenser (see [41]). In the simplest case of a two-layer condenser with specific resistances of the layers ρ_1 and ρ_2, dielectric constants ε_1 and ε_2, and layer thicknesses α_1 and α_2 respectively, when a voltage is applied instantaneously to the electrode an initial charge density of $+\sigma_0$ and $-\sigma_0$ arises. In this initial moment the strength of the field in the layers satisfies the condition

$$\frac{E_1^0}{E_2^0} = \frac{\varepsilon_2}{\varepsilon_1}. \tag{3}$$

However, in the following moment a leakage current arises which, generally speaking, is different for each layer because of the different resistances and different field strengths.

$$j_1 = \frac{E_1}{\rho_1}, \quad j_2 = \frac{E_2}{\rho_2}. \tag{4}$$

Because of the inequality of the conduction currents, charge accumulates at the boundary of separation of the layers, creating a field in consequence of which the conduction currents become equal. In the stationary state, the distribution of the field within the two-layer dielectric on the passage of a continuous current $j_1 = j_2$ satisfies the condition

$$\frac{E_1}{\rho_1} = \frac{E_2}{\rho_2}. \tag{5}$$

When an alternating current is passed through a two-layer condenser, a displacement current arises in it which is connected with the redistribution of the charge at the boundary of separation of the two phases and appears in the form of an additional capacitance of the condenser. The additional capacitance is determined from the formula

$$C_{\text{add}}=k\frac{\left(\dfrac{\varepsilon_1}{\rho_2}-\dfrac{\varepsilon_2}{\rho_1}\right)^2 d_1 d_2 S}{\left(\dfrac{d_2}{\rho_1}+\dfrac{d_1}{\rho_2}\right)^2(\varepsilon_1 d_2+\varepsilon_2 d_1)}, \tag{6}$$

where ε_1 and ε_2 are the dielectric constants and ρ_1 and ρ_2 are the specific resistances of the layers, d_1 and d_2 are the thicknesses of the layers, S is the area of a layer, and k is a constant.

It can be seen from the formula that the additional capacitance and, consequently, the effective dielectric constants are connected with the specific resistances of the layers and their thicknesses. An investigation of the change in the effective dielectric constant and specific resistance of a material under the influence of various factors can give valuable information on their submolecular structure.

The two-layer model can be complicated by the n-layer model [42] and by the model of a substance distributed in the form of spherical particles [43]. With further complication of the model, it is possible to consider the change in the concentration of current carriers at the boundary of separation of the two phases taking into account more accurately the influence of the diffusion of the carriers at the boundary of contact of the layers. As follows from the work of Trukhan [44], taking the space double layer at the boundary of separation of the phases into account leads to a lower value of the dispersion frequency. This error may prove substantial in the investigation of poorly conducting systems with dimensions comparable with the Debye screening radius. At frequencies exceeding the dispersion region, the conclusions of the theory agree with the conclusions of the Maxwell–Wagner theory [44]. This permits the approximate equations of the Maxwell–Wagner theory to be used in the plateau region.

Most frequently a model of one of the types considered is used which enables the observed phenomena to be described semiquantitatively. Such a model has been used in papers [29, 33, 38]. It is assumed that the substance mainly consists of two "sorts." The more highly conducting layers, or the regions of continuous conjugation, possess a specific resistance ρ_1 with an activation

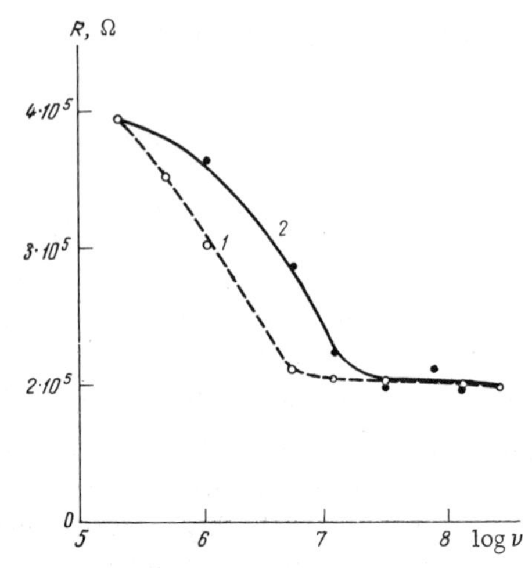

Fig. 15. Calculated (1) and experimental (2) relationships between the resistance of a film of polytetracyanoethylene and the frequency of the alternating current.

energy E_1. The substance of the thin layers separating the highly conducting layers possesses a specific resistance ρ_2 with activation energy E_2. The thickness of the barriers is considerably less than the thickness of the layers or grains of the main substance, and E_2 is greater than E_1. The dielectric constants of the two layers are assumed to be the same. In this case, $C_2 \gg C_1$, $R_2 \gg R_1$, and, consequently, $\tau_2 \gg \tau_1$, where C_1 and C_2 are the capacities of the main substance and the layers, R_1 and R_2 are the resistances of the main substance and the barriers, and the time constant $\tau = RC$. The capacity and resistance of such a system depend on the frequency. With an increase in the frequency, the barrier layers are shorted because of their high capacitance and the electrical properties of the main substance appear. At low frequencies, higher values of the dielectric constant must be observed because of the appearance of a charge at the grain boundaries.

In agreement with the model, films of a polymeric complex of tetracyanoethylene (TCE) on metals exhibit a high dielectric constant. Thus, for example, for films on iron its magnitude may be 50-70 and on aluminum it reaches 6000 at a frequency of several kHz [32]. As the model requires, the effective dielectric constant decreases with an increase in the frequency of the alternating current.

TABLE 8. Correlation between the Pre-exponential Factor and the Specific Resistance Obtained with Alternating Current

Sample	ρ_0	ρ_ν
	$\Omega \cdot cm$	
Metalfree polytetracyanoethylene	$4.76 \cdot 10^3$	$1.12 \cdot 10^3$
Complex of TCE with silver	1.78	1.41
Modified polyethylene treated at 620°C	$2 \cdot 10^3$	$2.5 \cdot 10^3$
Modified polyethylene treated at 420°C	$3.3 \cdot 10^3$	$2 \cdot 10^3$

According to the model, with a direct current and at low frequencies it is mainly the resistance of the gap or the layers between the regions of continuous conjugation that is measured. With an increase in the frequency, the layers are shorted because of the high-capacitance conduction and the capacitance and resistance of the regions with a lower resistance and capacitance begin to appear. Figure 15 shows the resistance of a film of polytetracyanoethylene as a function of the frequency in comparison with the calculated curve.

If the model with a heterogeneous structure is correct and at a high frequency in the region of the plateau it is actually the resistance of the regions of continuous conjugation in which the electrons move freely that is measured, then the magnitude ρ_0 from the formula

$$\rho = \rho_0 \exp\left(\frac{\Delta E}{kT}\right)$$

and the magnitude ρ_ν measured at a high frequency in the region of the plateau should coincide. Table 8 gives the results of such a comparison for samples of polyethylene modified by radiation and thermal methods and for samples of polytetracyanoethylene [40].

On the basis of the model considered, it was to be expected that the activation energy of conduction should decrease with an increase in the frequency of the alternating current, when the parameters of the region of continuous conjugation are measured. Figure 16 shows the temperature dependence of the thermal activation energy for films of polytetracyanoethylene in the range of frequencies from 0 to 200 MHz. As can be seen from the figure, at a frequency of 200 MHz, the activation energy is close to zero. Similar results have been obtained for radiation-thermally modified polyethylene [40].

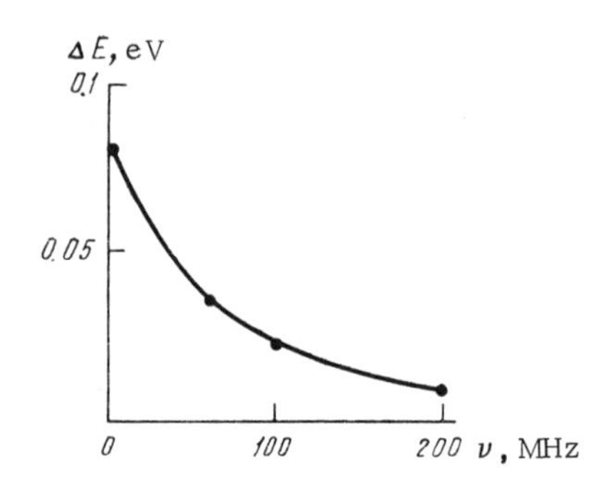

Fig. 16. Thermal activation energy of conduction for a polytetracyanoethylene film as a function of the frequency of the current.

A constant electric field is concentrated mainly between the highly conducting sectors, which must be shown in the volt–ampere characteristics. In actual fact, in a study of films of organic semiconductors prepared from polyethylene [34], it was shown that the i–v curves have the following form: in the region of low fields ($\leqslant 10^3$ V/cm) Ohm's law is obeyed, while at higher field strengths the current rises exponentially with an increase in the applied voltage. Figure 17 shows the volt–ampere characteristics for one of the samples on a double logarithmic scale at various temperatures. Let us consider the possible mechanisms responsible for the nonlinearity in the volt–ampere characteristics. Tunneling by the electrons of the film can be excluded immediately, since at the thicknesses studied (up to 10 μ) this mechanism is unlikely. Schottky emission and currents limited by the space charge cannot appear in this case, since the effect does not depend on the material of the electrodes. It is possible to exclude the trivial cause of nonlinearity connected with the heating of the sample by Joule heat.

For the nonhomogeneous structure considered above, the barrier mechanism of conduction is apparently applicable. In order to obtain an expression for the macroscopic conductivity it is possible to neglect the resistance of the regions of polyconjugation and write the current density as a function of the applied voltage passing through the barrier [45]:

$$j = Mep\left[\exp\left(-\frac{\varphi}{kT}\right)\right]\left[\exp\left(\frac{e\Delta v}{kT}\right)-1\right], \tag{7}$$

where p is the mean concentration of the majority carriers (holes) in the regions of polycondensation, φ is the height of the potential

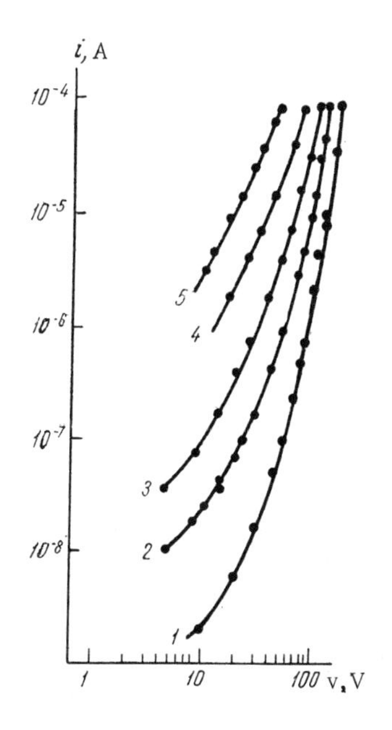

Fig. 17. Volt–ampere characteristics of a sample of modified polyethylene. 1) −17°C; 2) 20°C; 3) 50°C; 4) 93°C; 5) 137°C.

barrier, Δv is the voltage drop at one barrier, and M is a parameter not depending on φ and being of the order of the thermal velocity of the carriers. In the region of weak fields, where $\Delta v \gg kT/e$, equation (7) is transformed into the following equation:

$$j = \frac{Mpe^2\Delta v}{kT} \exp\left(-\frac{\varphi}{kT}\right). \tag{8}$$

Assuming that there are N successively connected regions of polyconjugation, it is possible to define Δv

$$\Delta v = \frac{v}{N} = \frac{v}{nd} = \frac{E}{n}, \tag{9}$$

where v is the voltage applied to the sample, E is the field strength, d is the thickness of the sample, and n is the number of regions of polyconjugation per unit length. In the region of strong fields, $\Delta v \gg kT/e$, from (7) and (6) we obtain

$$j = Mep \exp\left(-\frac{\varphi - \frac{e}{nd} v}{kT}\right) = Mep \exp\left(-\frac{\Delta E}{kT}\right), \tag{10}$$

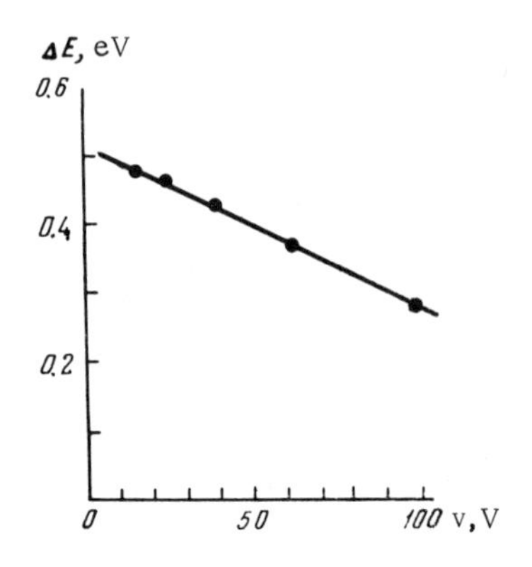

Fig. 18. Relationship between ΔE and v obtained from the data given in Fig. 17.

where $\Delta E = \varphi - Bv$, $B = e/nd$. Curve 2 in Fig. 17, corresponding to measurements at room temperature, was plotted from the equation

$$j = 4.8 \times 10^{-8} \left[\exp\left(\frac{1.6 \times 10^{-3} v}{kT} \right) - 1 \right],$$

from which $B = 1.6 \times 10^{-3}$.

We must recall yet another space effect leading to nonlinear volt–ampere characteristics, namely, the Frenkel'–Poole effect

$$j = G_0 E \exp\left(-\frac{\varphi' - B' v^{1/2}}{kT} \right), \tag{11}$$

where G_0 is a constant having the dimensions of conduction, φ' is the ionization energy of the impurity atoms,

$$B' = \left(\frac{e^3}{\pi \varepsilon d} \right)^{1/2}; \tag{12}$$

and ε is the dielectric constant.

If one approximates the volt–ampere characteristic to a relation of type (11), similar values of B' are obtained: the value calculated from (12) and the value obtained from the curve. However, the choice between the barrier mechanism and the Frenkel'–Poole effect can be made from the type of dependence of ΔE on v in the region of strong fields. In actual fact, in the first case ΔE changes linearly with v and in the second case linearly with $v^{1/2}$. Figure 18 gives the activation energies of conduction obtained from the curves of Fig. 17 as a function of the potential difference applied to the sample. As follows from the figure, ΔE changes linearly with a change in v, $\Delta E = 0.51 - 2.10 \times 10^{-3}$ v, i.e., $B = 2.1 \times 10^{-3}$, in good agreement with the value of B obtained from the volt–ampere characteristic at room temperature and given above.

On the basis of the value of B obtained it is possible to determine the number of regions of polyconjugation per unit length, which proves to be 2.5×10^6 cm^{-1}. Assuming that the regions of polyconjugation occupy the bulk of the volume of the substance, we find the linear dimension of one region: $L \approx 40$ Å. With a rise in the temperature of treatment of the samples, the conductivity at room temperature rises and the coefficient B in expression (10) increases as a result of the increase in the size of the highly conducting segments and an increase in the linear dimensions, reaching ~300 Å according to the experimental results.

Taking into account the fact that $\sigma = j/E$, we can write equation (8) in the following form:

$$\sigma = ep \frac{Me}{kTn} \exp\left(-\frac{\varphi}{kT}\right). \tag{13}$$

Here the member after p may be considered as the effective mobility:

$$\mu^* = \frac{Me}{kTn} \exp\left(-\frac{\varphi}{kT}\right). \tag{14}$$

The concentration of carriers p may also vary with a change in the temperature, but it is important to stress that transitions from one region to another determine the temperature dependence of the effective mobility.

Within the limits of the regions of polyconjugated bonds, the band model is applicable. By defining the mobility of the carriers in the regions as μ_c and introducing the relaxation time τ_r, it is possible to define the length of the free path $l = \tau_r v_{th}$ where v_{th} is the thermal velocity of the charge carriers.

In the simplest case, this leads to the following expressions for μ^* and σ:

$$\mu^* = A\mu_c \frac{L}{l} \exp\left(-\frac{\varphi}{kT}\right), \tag{15}$$

$$\sigma = A\sigma_c \frac{L}{l} \exp\left(-\frac{\varphi}{kT}\right), \tag{16}$$

where σ_c is the conductivity of the regions of polyconjugated bonds and A is close to unity in order of magnitude. Interesting results from this point of view have been obtained in measurements of the current noise in pyrolyzed polyacrylonitrile [48]. It was found that the considerable amplitude of the noise current can be explained by the fact that the specimen consisted of a large number of highly-conducting subvolumes the concentration of which $\leq 10^{17}$ for samples with a conductivity of $10^{-3}\,\Omega^{-1}\cdot\text{cm}^{-1}$. This means that each region consists of $\geq 5 \times 10^5$ carbon atoms. In better conducting samples ($\sigma \sim 10^{-2}\,\Omega\cdot\text{cm}^{-1}$), the regions of polyconjugation increase in dimensions and contain as many as $\geq 5 \times 10^8$ carbon atoms. A study of the frequency spectrum of the current noise gives grounds for concluding that the basic structure of the regions is preserved as they become larger.

It should be noted particularly that the observed heterogeneous structure is not merely a trivial consequence of the imperfection of the material, as may appear at first sight. When the dimensions of the molecules already composing an ideally ordered system approximate those of a subphase, the role of the contact phenomenon due to the "surface" of the macromolecule increases. This may appear in the marked dependence of the electrical properties on adsorbed gases and also in the frequency characteristics of such systems. Generally, the properties of a material consisting of layers or granules of subphase dimensions must differ from the properties of the same substance in bulk because of the double layers formed at the boundary of separation of the phases.

On the application of the field, the carriers can move only within the limits of the regions of continuous conjugation. Then polarization of that part of the medium where these movements are restricted takes place. The transfer of charge between the regions of continuous conjugation at high temperatures obviously takes place by an activation mechanism. The shorter macromolecules are electron acceptors in relation to the longer ones as follows from data on the change in the work function of an electron when certain polymeric semiconductors are heated [39, 40].

Thus, we have a substance consisting completely of "impurities" of both donor and acceptor type. The peculiarity of such "impurities" consists in the fact that they themselves possess a large number of carriers taking part in conduction which, however, can be detected in measurements with alternating current. The indeterminacy of the composition of polymeric semiconductors is the main difficulty in interpreting the electrical measurements.

DIRECT MEASUREMENT OF THE MOBILITY OF THE CURRENT CARRIERS

To measure the mobility of the current carriers in organic semiconductors the most powerful and promising method is that which was well developed originally for crystal lattices of charged particles and consists in measuring the drift of excited carriers in an electric field. As applied to organic semiconductors, this method can be used to measure mobility both in single crystals and in polycrystalline layers of low-molecular-weight organic semiconductors and in thin films of polymeric semiconductors. The mobility may be extremely small (approximately from ~10 to 10^{-6} $cm^2/V \cdot sec$) and the dark conductivities must be low. Organic semiconductors satisfy these conditions in the best possible way.

In view of the great importance of this method, we shall consider it in more detail. Figure 19 shows the process occurring in a layer of semiconductor on excitation and gives the equivalent scheme of the switching of the sample. The injection of nonequilibrium carriers can be carried out by photons or low-energy electrons (1-20 keV) and also from an ohmic contact with the switching on of the electric field. Photoexcitation was first considered in anthracene [49, 50]. Here the top electrode (see Fig. 19) is semi-

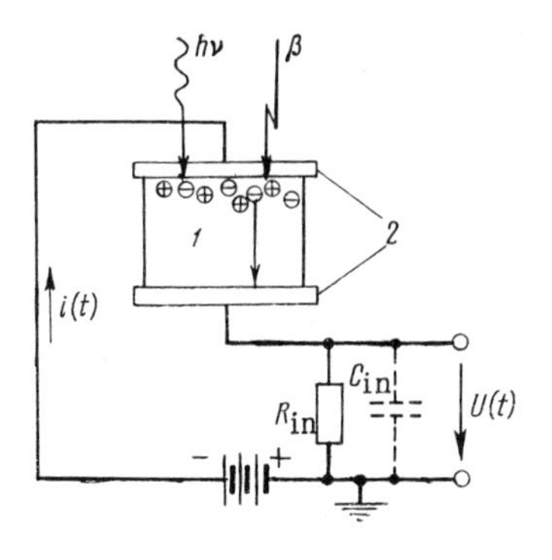

Fig. 19. Scheme for recording the movement of injected carriers in a sample. 1) Sample; 2) electrodes.

transparent, and the luminous flux is absorbed in the thin electrode layer. In this case the material studied must be photoconductive. If the sample is not photoconductive within a given region of the spectrum, it is possible to use electron bombardment [51]. The layer of the absorption of low-energy electrons usually does not exceed fractions of a micron.

With electronic excitation it is easily possible to control the intensity, length, and frequency of the electronic pulses, and a contraction or expansion of the ionization layer is achieved by changing the energy of the exciting electrons. Furthermore, in this case it is fairly simple to measure the number of electrons of a given energy in a pulse. By measuring the number of carriers formed it is possible to evaluate the energy required for the formation of one electron–hole pair, which is not always possible in photoexcitation. If the electrode injects carriers of one sign into the sample, the drift mobility can be measured by the same arrangement without external sources of excitation. For this we must apply to the specimens voltage pulses the growth time of which is far less than the time of transfer of the carriers to the second electrode [52].

In the measurements, a constant attractive field is applied to the specimen. At moment of time $t = 0$ nonequilibrium current carriers are excited at the upper electrode (by electrons or photons). Depending on the polarity of the attractive field, a signal is recorded from the movement of the holes or electrons to the second electrode; the current through the sample ceases when the carriers reach the electrode after transfer time T_t. The drift mobility is determined from the relation

$$\mu = \frac{d}{ET_t}, \tag{17}$$

where d is the thickness of the sample and E is the electric field strength. With an increase in the field strength the transfer time decreases. This shows that the carriers reach the collector electrode and, consequently, the mobility is determined from formula (17). If the sample contains deep traps with a trapping time far greater than the transfer time, the carriers fall into the traps and do not reach the second electrode. In this case, with an increase in the field strength the duration of a current pulse due to the motion of the carriers remains constant but increases in amplitude,

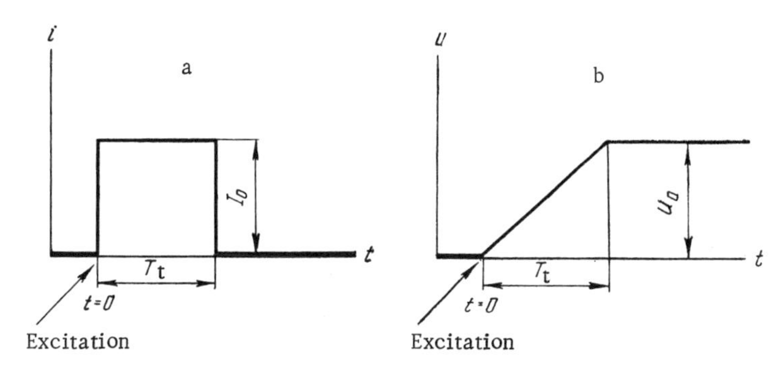

Fig. 20. (a) Current pulse in the sample and (b) voltage pulse at the output of the circuit for the case $R_{in} = \infty$.

and determines the life of the carriers before their fall into the deep traps. If the carriers reach the collector electrode, then, as a rule, a linear relationship is observed between E and $1/T_t$. This shows that the mobility does not depend on the field strength. Certain conditions exist under which it is possible to measure the mobility by the method considered [53]: a) the pairs of charged carriers must be generated after time t_g, which is far less than the transfer time T_t; b) the generation layer λ must be far smaller than the thickness of the sample d; c) the attractive electrical field in the sample must be rigidly uniform; d) the time of dielectric reaction (in essence the constant time $R_{spec} \cdot C_{spec}$) t_r, $10^{-12}\ \varepsilon/4\pi\sigma$ (ε and σ in practical units), must be far greater than the transfer time; and e) the electrodes must not inject electrons.

In the ideal case, when there are no traps for the current carriers and they all reach the second electrode under the action of the exciting pulse, a charge q appears on the collecting electrode. This corresponds to a current pulse i(t), which is fed into a circuit consisting of R_{in} and C_{in}. The change in the current pulse with time is shown in Fig. 20a. In the general case, q is given by the formula $q = \int_0^\infty i(t)\,dt$, and in the given case $q = I_0 T_t$, where I_0 is the amplitude of the current pulse.

The voltage at the output of the circuit U(t) is written in the general case as

$$U(t)_{0-T_t} = U_0 \frac{R_{in} \cdot C_{in}}{T_t}\left(1 - e^{-\frac{t}{R_{in} \cdot C_{in}}}\right) \tag{18}$$

$$U(t)_{T_t-\infty} = U_0 \frac{R_{in} \cdot C_{in}}{T_t}\left(e^{-\frac{(t-T_t)}{R_{in} \cdot C_{in}}} - e^{-\frac{t}{R_{in} \cdot C_{in}}}\right), \tag{19}$$

where $U_0 = (I_0 \cdot T_t)/C_{in}$, and the maximum amplitude U_{max} of the voltage U(t) which is reached at $t = T_t$ is

$$U_{max} = U_0 \frac{R_{in} \cdot C_{in}}{T_t}\left(1 - e^{-\frac{T_t}{R_{in} \cdot C_{in}}}\right). \tag{20}$$

In this case, when the measurements are carried out in thin single crystals and it is possible to achieve a minimum value of C_{in}, in which C_{spec} is included, the condition $R_{in}C_{in} \ll T_t$ is satisfied [49, 54-57]. In measuring samples of thin-film polycrystalline and polymeric organic semiconductors with a high capacity, the condition $R_{in}C_{in} \gg T_t$ is satisfied [51, 58-63].

It is easy to see that in the first case formulas (18)-(20) assume the form:

$$U(t)_{0-T_t} = I_0 \cdot R_{in}, \tag{18'}$$

$$U(t)_{T_t-\infty} = 0, \tag{19'}$$

$$U_{max} = I_0 R_{in}, \tag{20'}$$

and in the second case (Fig. 20b)

$$U(t)_{0-T_t} = U_0 \frac{t}{T_t}, \tag{18''}$$

$$U(t)_{T_t-\infty} = U_0, \tag{19''}$$

$$U_{max} = U_0. \tag{20''}$$

Thus, in the first case a pulse of current is measured and in the second case the voltage arising in the capacity C_{in} on the passage of the current carriers through the sample. In actual fact, any sample studied contains trapping levels into which a current carrier falls and during a short (in comparison with T_t) time does not participate in the movement through the specimen. Moreover, deep traps are possible (particularly in the case of polycrystalline specimens) in which the time of residence in them τ_{tr} may be comparable with T_t or even greater.

In its general form, an analysis of the motion of the generated carriers in a crystal with traps has been given by Thornber and Mead [64]. From a consideration of the time dependences of the free charges and those caught in traps, the following expression was obtained for the charge on the electrode arising on the drift of carriers through the crystal:

$$q(t)=\frac{n_0 e}{T_t}\left[\frac{\tau_0}{\tau_{tr}}\cdot t+\frac{\tau_0^2}{\tau}\left(1-e^{-\frac{t}{\tau_0}}\right)\right], \tag{21}$$

where τ is the lifetime of the carriers before their capture by deep traps, n_0 is the number of carriers generated per impulse, and $1/\tau_0 = (1/\tau) + (1/\tau_{tr})$. The magnitude of the free charge participating in transfer is expressed by the following equation

$$q_{free}(t)=n_0 e\left(\frac{\tau_0}{\tau_{tr}}+\frac{\tau_0}{\tau}e^{-\frac{t}{\tau_0}}\right), \tag{22}$$

and the magnitude of the total charge captured by the deep traps and not participating in conduction

$$q_{tr}(t)=n_0 e-q_{free}(t). \tag{23}$$

Then between q(t) and q_{free} (t) the following relation exists:

$$q(t)=\frac{1}{T_t}\int_0^t q_{free}(t)\,dt. \tag{24}$$

With a very long time of residence of the charge carriers in the trap τ_{tr}, expression (21), when (17) is taken into account, assumes the well-known form

$$q(t)=\frac{n_0 eE\mu\tau}{d}\left(1-e^{\frac{-t}{\tau}}\right) \tag{25}$$

Let us consider some limiting cases that are most frequently encountered in measurements.

1) $T_t \ll \tau \ll \tau_{tr}$, i.e., in spite of the presence of deep traps the carriers are not captured by them ($T_t \ll \tau$) and therefore even a large trapping time ($\tau_{tr} \gg \tau$) has no effect on the shape of the charging curve. Then

$$q(t)\approx\frac{n_0 e}{T_t}\cdot t \quad \text{for} \quad t\leqslant T_t;$$

$$q(t)\approx n_0 e \quad \text{for} \quad t>T_t.$$

The induced charge rises linearly with its original rate to the maximum value.

2) $\tau \ll T_t \ll \tau_{tr}$; the lifetime of the carriers before they fall into deep traps is far smaller than the time of transfer of the charge carriers ($\tau \ll T_t$). The carriers that have fallen into the traps no longer participate in the transfer of charges through the specimen ($\tau_{tr} \gg T_t$). In this case, we obtain an expression identical with equation (25).

$$q(t) = n_0 e \left(\frac{\tau}{T_t}\right)\left(1 - e^{-\frac{t}{\tau}}\right) \quad \text{for} \quad t \leqslant T_t;$$

$$q(t) = n_0 e \left(\frac{\tau}{T_t}\right)\left(1 - e^{-\frac{T_t}{\tau}}\right) \quad \text{for} \quad t > T_t.$$

The linear part is far shorter, and considerably fewer charge carriers reach the second electrode than in case one.

3) $\tau_{tr} \ll \tau \ll T_t$; this means that during transfer the charges repeatedly fall into traps and are rapidly liberated.

4) $\tau_{tr} \ll T_t \ll \tau$; under this condition, in spite of the presence of traps the carriers do not fall into them during transfer ($T_t \ll \tau$). Two subsequent cases may be regarded as effects of shallow traps:

$$q(t) \approx \frac{n_0 e}{T_t} \cdot t \left(1 + \frac{\tau_{tr}}{\tau}\right) \approx \frac{n_0 e}{T_t} \cdot t \quad \text{for} \quad t \leqslant T_t;$$

$$q(t) \approx n_0 e \qquad \text{for} \quad t > T_t.$$

It follows from this that if the time of capture by the traps is far smaller than the time of transfer, then, regardless of the lifetime of the carriers before they fall into the traps, the charge–time characteristics have the same nature as in the case of a long residence time in a trap at a time of transfer far smaller than the life of a carrier. However, it does not follow from this that the mobilities will be the same in these cases, since the time of transfer T_t from which the drift mobility is calculated will be greater when shallow traps are present than when they are absent at the same field strength and specimen thickness.

MOBILITY OF CURRENT CARRIERS IN MOLECULAR CRYSTALS

Let us now consider the experimental results obtained in measurements of the characteristics of the transfer of charge carriers in organic semiconductors. The first and, up to the present time, most complete characteristics have been obtained in a study of monocrystals of anthracene.

In a study of the movement of photoexcited holes, it was found that their mobility is between ~ 0.5 and $1.5\ cm^2/V \cdot sec$ [49, 50, 65]. These results have been completely confirmed in further investigations [52, 59, 63].

The temperature dependence of the mobility in all cases has the form $\mu \sim T^{-n}$ but the value of n varies in measurements on different samples from 1 to 2.3. It may be mentioned that in the case of scattering on acoustic phonons n must be 1.5. The trailing edge of the pulse has a prolonged fall the length of which is considerably greater than could be explained by the diffusion of the carriers in the layer of space charge.

At the same time, the observed temperature dependence of the mobility does not permit an explanation of the diffuseness of the edge by the existence of traps capturing the holes [cf. equation (25)] since in this case an activation rise in the mobility with an increase in the temperature should be observed. A more detailed study of monocrystals of anthracene gives grounds for assuming that in addition to the release of excited holes into the bulk the process of capture of part of the holes by surface states takes place [59].

This assumption is the more probable because the first stage in the formation of photoexcited free charge carriers in anthracene is the formation of a molecular exciton, which is localized on surface defects. In the thermal decay of the exciton, an electron is apparently trapped by a defect and a hole passes into the bulk of the crystal. The energy necessary for this process is 0.14 eV [59]. In the thermal emptying of these states, the holes make a contribution to the observed current pulse. The lifetime of the holes is $\sim 10^{-4}$ sec and in very pure crystals it increases to 2×10^{-3} sec [55]. The possibility of another cause for the prolonged fall in the current pulse has been given by other authors [52, 66]. In the movement of the charge carriers through the crystal, space charge limited current conditions which occur where there is a considerable "reservoir" of carriers from which they are partially drawn off by the electric field, may arise. In ohmic contact the electrode is such a "reservoir." The possibility of obtaining an ohmic contact in organic crystals has been considered by Gränacher [67].

With high intensities of the existing light, if we can neglect the processes of recombination and decay, the carriers generated at one of the electrodes form a so-called virtual cathode close to the illuminated surface of the crystal.

If the motion of the charge carriers is limited by the space charge, an extended tail is observed even in the absence of trapping states. The time of transfer determined from the form of the current pulse is 20% less than the true time of transfer T_t in a homogeneous electric field $E = V/d$, since the total field is increased through the space charge. If trapping levels exist in the crystal, the difference becomes still smaller.

In order to exclude the effects of the space charge, the measurements must be carried out at high field strengths and low concentrations of generated carriers. However, if the current is limited by the space charge and this is not allowed for in the calculation of mobility, the error in the values of μ does not exceed 20%, which is completely acceptable in such measurements.

The results for anthracene show that in that case the microscopic mobility of the holes is measured. The characteristics of the mobility of the electrons are sensitive to the perfection of the crystals and the lifetime of the electrons varies approximately from 2×10^{-7} sec in crystals not subjected to special purification [63] to 2×10^{-3} sec in very pure crystals [55]. It has been shown [49, 65] that the mobility of the electrons has the same temperature dependence as the mobility of the holes, and the microscopic mobility, like that of the holes, is approximately 1 $cm^2/V \cdot sec$. The observed temperature dependence of the mobilities shows the applicability of the band theory. With scattering on the thermal vibrations of the lattice, $\mu \sim \frac{T^{-3/2}}{m^{*x^{5/2}}}$, where m^* is the effective mass of the carriers. The mobility strictly depends on the effective mass, and in its turn the effective mass must be sensitive to

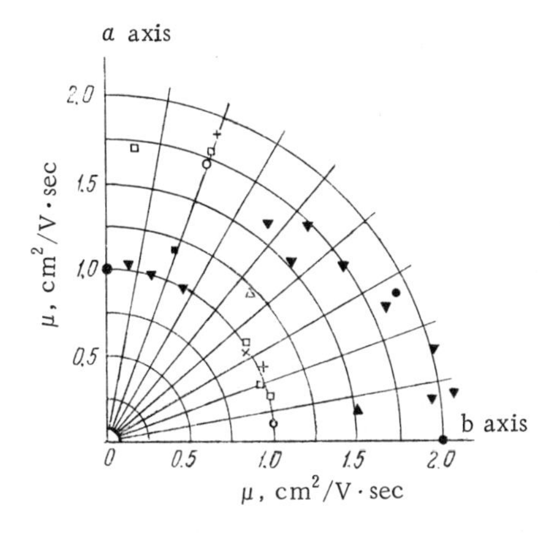

Fig. 21. Mobility of electrons and holes as a function of the orientation of anthracene crystals. Black symbols, electrons; white symbols, holes.

the lattice constants, particularly in a molecular crystal. In connection with this, it was interesting to trace the influence of the pressure on the mobility of charge carriers.

The quantitative discussion given by Le Blanc [68] shows that with a decrease in the volume of the crystal by 3%, as has been observed in experiments [65], the width of the bands both for electrons and for holes increases by approximately 20%. To a first approximation, the mobilities must increase in the same ratio. The calculated increase in mobility is somewhat less than that determined experimentally, but it is of the same order of magnitude.

In a calculation of the band structure of anthracene [69] it was shown than an anisotropy of the mobilities must be observed, this being determined by the structure of the crystal and the symmetry of the molecular wave functions. The calculation of the anisotropy was carried out for two cases of scattering: with free paths of constant time, on the one hand, and of constant length, on the other. The results of the experimental determination of the mobility as a function of the selected direction in the crystals are given in Fig. 21 [65]. The mobility of the holes is greater along the *b* axis than along the *a* axis, and the situation is the opposite for the electrons, which agrees quantitatively with the calculations performed.

As already mentioned, the electrons in a crystal of anthracene are more sensitive to a given type of disturbance than the holes. As early as 1960 [49] it was established that on photoexcitation ten times fewer electrons than holes participate in the transfer of charge carriers through the sample. If a crystal of anthracene is not subjected to special purification, the magnitude of the signal from the motion of the electrons is below the sensitivity of the apparatus [59]. The hypothesis has been put forward [70] that molecules of oxygen or its compounds included during the growth of the crystal may act as effective traps for electrons. In particular, molecules of anthraquinone may form such an impurity.

Hoesterey and Letston [55] doped very pure crystals of anthracene (lifetime of electrons $\sim 2 \times 10^{-3}$ sec) with a known amount of anthraquinone. The signal from the motion of the electrons practically disappeared. It would seem that it is not possible to measure the characteristics of the traps and the transfer of electrons by the method considered. Nevertheless, such measurements could be carried out by using the method proposed by Adams and Spear [71].

Such an investigation has been carried out for monocrystals of anthracene not subjected to special purification in an investigation [63] in which low-energy electrons were used to generate the carriers. Intensive irradiation of the surface of the crystal was carried out with the application of an attractive field favoring the passage of electrons from the region of generation into the sample. The capture of the electrons by deep traps took place with the formation of a space charge in the immediate vicinity of the region of generation. The lifetime of the electrons before their capture by the deep traps did not exceed 2×10^{-7} sec, while for holes it was $3\text{-}5 \times 10^{-5}$ sec. Then, in different intervals of time greater than the time of irradiation the specimen was probed by a short exciting pulse with zero external voltage on the specimen. Pulses due to the motion of the holes retained in the bulk of the crystal by the field of the space charge of the trapped electrons were observed. Thus, the amplitude of the pulse of the holes is a measure of the space charge. As can be seen from Fig. 22, the amplitude of the pulse due to the holes after the formation of a space charge decreases by an exponential law and

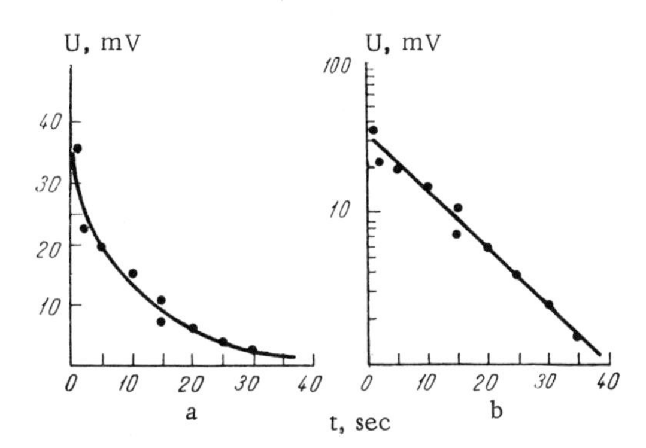

Fig. 22. Decrease in the space charge of electrons with time in anthracene in linear (a) and semilogarithmic (b) coordinates.

$$q_1(t) = q_1(0) \exp\left(-\frac{t}{\tau_{tr}}\right),$$

where τ_{tr} is the mean time of residence of the electrons in the traps. At 20°C, $\tau \approx 12$ sec.

From temperature measurements, the depth of the traps was estimated as 0.6 eV. In the simplest case, $\tau_{tr} = (N_s v\sigma)^{-1} \exp\left(\frac{\Delta E}{kT}\right)$, where v is the thermal velocity of the charge carriers, N_s is the effective density of the states and σ is the trapping cross section. Calculations of the band structure of anthracene [69, 72] have shown that at ordinary temperatures the widths of the conduction band < kT and consequently N has the order of concentration of the molecules and the mean thermal velocity of the electrons is 10^5 cm/sec. From this we obtain for σ a value of 10^{-17} cm^2. Subsequent measurements showed that the concentration of electrons traps N_{tr} in these crystals is greater than 5×10^{16} cm^{-3} and the microscopic mobility of the electrons $\mu > 0.3$ cm^2/V·sec. This agrees well with the measurements of other authors [49, 50, 59].

In spite of the fact that holes are less sensitive to imperfections of the anthracene crystal, it is possible to select impurities which are traps for holes. It has been shown [55] that such traps can be formed by molecules of naphthacene introduced in a definite concentration into a single crystal of anthracene. This is accompanied by an increase in the time of transfer T_t of the holes, which

TABLE 9. The Relation between Drift and Microscopic Mobilities in Anthracene Doped with Naphthacene

Concentration of impurities, parts per million	$\mu_{dr}/\mu \cdot 10^2$ (at 300°K)	
	found	calculated
1	7.1	6.3
2	5.0	3.1
5	1.3	1.3
10	0.71	0.63
40	0.54	0.16

determines their mobility, from 50 μsec in pure anthracene to 1000 μsec in the doped material, i.e., the mobility decreases by more than thirty times. The mobility in the doped anthracene has an activation dependence on the temperature, $\mu_{dr} = \mu \exp\left(-\frac{\Delta E}{kT}\right)$, where $\Delta E = 0.43$ eV. The results obtained can be explained if it is assumed that the holes repeatedly fall into traps with a depth of 0.43 eV. In this case, the drift mobility is connected with the microscopic relation

$$\mu_{dr} = \mu \frac{\tau}{\tau + \tau_{tr}}. \tag{26}$$

In its turn

$$\frac{\tau}{\tau_{tr}} = \frac{N_s}{N_{tr}} \exp\left(-\frac{\Delta E}{kT}\right). \tag{27}$$

At low temperatures, while $\tau \ll \tau_{tr}$, μ_{dr} changes exponentially with a change in the reciprocal temperature. At high temperatures $\tau > \tau_{tr}$ and the drift mobility reaches the microscopic mobility in magnitude and approximates its temperature dependence. Assuming that each molecule of naphthacene is a trap for holes, from equations (26) and (27) we can calculate the ratio η_{dr}/μ in the region $\tau \ll \tau_{tr}$ $\left(\frac{\mu_{dr}}{\mu} = \frac{\tau}{\tau_{tr}}\right)$, assuming that N_{tr} is equal to the concentration of naphthacene and N_s is equal to the density of the anthracene molecules. Table 9 gives calculated and experimental results.

The good agreement between the figures found and calculated confirms the proposed mechanism of the influence of naphthacene on the transfer of holes.

In investigations on the mechanism of the transfer of carriers in other organic semiconductors, other laws have been found in addition to those first established in the study of anthracene. Results very similar to those obtained in the work on anthracene have been obtained in a study of charge transfer in a monocrystal of pyrene [73]. The lifetime of the holes in pyrene exceeds the lifetime of the electrons and the mobilities of the carriers are equal at 0.75 $cm^2/V \cdot sec$ in the direction parallel to the *ab* plane of the monoclinic crystal. When different directions in the crystal are selected, anisotropy of motion is observed and the temperature dependence of the mobilities of the holes and the electrons obeys the T^{-n} law. It follows from this that the mechanism of the mobility in monocrystals of pyrene is the same as in anthracene.

In a study of monocrystals of ferrocene, triphenylamine, and trans-stilbene it was found that the mobilities of the holes of these substances depend on the temperature in a different manner, as is shown in Fig. 23 [56]. For comparison, the figure also gives the temperature dependence of the mobility of the holes in anthracene. In the substances considered the signal from the motion of the photoexcited electrons was small and the mobility could not be measured, since τ was less than 1 μsec.

In trans-stilbene the small slope of the plot of the mobility as a function of the temperature, which is close to that for anthracene, and the mobility, which is 2.4 $cm^2/V \cdot sec$ at room temperature, validate the conclusions that were drawn in the study of anthracene and, consequently, confirm the applicability of a band conduction mechanism.

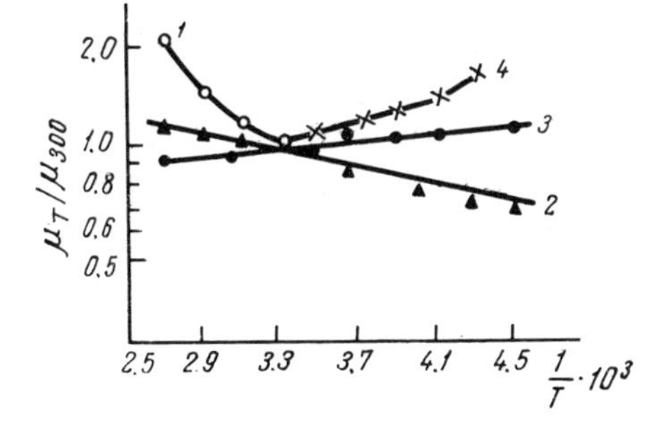

Fig. 23. Temperature dependence of the mobility of holes for various substances. 1) Triphenylamine; 2) ferrocene; 3) trans-stilbene; 4) anthracene.

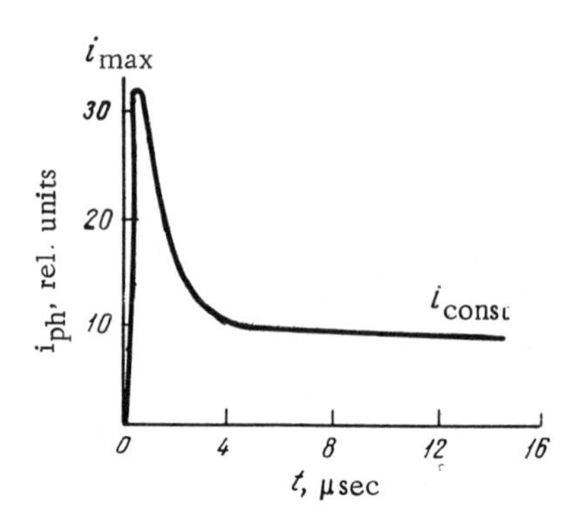

Fig. 24. Oscillogram of the photocurrent in a crystal of triphenylamine. Length of the flash of light 0.2 μsec.

In triphenylamine the change in mobility with a change in the temperature gives an activation energy of 0.5 eV and, consequently, the transfer time is determined by the drift mobility. In order to obtain the microscopic mobilities, which are given in Fig. 23, we could use equations (26) and (27). However, in the triphenylamine crystals the values of N_s and N_{tr} are unknown. But the microscopic mobility may nevertheless be obtained by investigating the behavior of the photocurrent immediately after the pulse photogeneration of carriers. As shown in Fig. 24, the photocurrent reaches a maximum value i_{max} immediately after the process of generation and then falls exponentially to a constant value i_{const} and does not change in time T_t during which the carriers move through the crystal. This behavior of the photocurrent can be understood by considering equations (22)-(24). The photocurrent registered in the circuit is

$$\frac{dq\,(t)}{dt}=\frac{q\,(t)}{T_t}=\frac{n_0 e\cdot\tau_0}{T_t\cdot\tau_{tr}}\left(1+\frac{\tau_{tr}}{\tau}\,e^{-\frac{t}{\tau_0}}\right). \tag{28}$$

In the region of purely activation mobility, $\tau_{tr} \gg \tau$ and $\tau_0 = \tau$. Immediately after the regeneration process, t is very small and $i_{max} = n_0 e/T_t$, i.e., all the carriers formed, before falling into traps, are free and make a contribution to the photocurrent. Then a decrease in the photocurrent follows an exponential law with the time constant τ. At the stationary concentration of free carriers

$$i_{const}=\frac{n_0 e}{T_t}\cdot\frac{\tau}{\tau_{tr}}.$$

Thus, taking (27) into account we obtain:

$$\frac{i_{\text{const}}}{i_{\text{max}}} = \frac{\tau}{\tau_{\text{tr}}} = \frac{N_s}{N_{\text{tr}}} \exp\left(-\frac{\Delta E}{kT}\right) = \frac{\mu_{\text{dr}}}{\mu}. \tag{29}$$

By measuring i_{const}, i_{max}, and μ_{dr} it is possible to calculate the microscopic mobility μ, which is given in Fig. 23. On the basis of these measurements we can also calculate the concentration of traps, which determines the drift mobility, and the trapping cross section of the carriers, which are $\sim 10^{13}$ cm^{-3} and $\sim 10^{-14}$ cm^2, respectively.

Organic semiconductors cannot always be obtained in the form of monocrystals. Far more frequently they are obtained in the form of polycrystalline films of different thicknesses. In this case, the process of transfer is greatly complicated since the boundaries of the crystallites create additional hindrances to the motion of the charge carriers.

In an investigation of polycrystalline films of p-terphenyl [60] it was shown that the amplitude of the voltage pulses increases linearly with an increase in the attracting field. This shows that only part of the carriers generated in the ionization layer reach the collector electrode. The trailing edge of the pulse shows the presence in the sample of deep traps decreasing the concentration of carriers n with time. An estimate of the lifetimes of the holes and the electrons gave values of 3 and 2.4 μsec, respectively, while T_t amounted to several tens of microseconds. It would appear that here we are dealing with case 3 (see p. 84); however, for these specimens a more complex relationship not reducible to the limiting cases considered is observed, namely, $\tau \ll \tau_{\text{tr}} \ll T_t$, i.e., in penetrating the film the carriers repeatedly fall into the traps with a long time of capture. In addition to these traps, the authors also assume the existence of deep traps the capture of which is considerably greater than T_t. By considering such a model, the following expressions were obtained for the voltage pulses:

$$U(t, v) = \frac{e \cdot n_0 v}{C_{\text{in}} \cdot d^2} \left\{ \mu_1 \tau \left(1 - e^{-\frac{t}{\tau}}\right) + \mu_2 \frac{\tau_2 \cdot \tau_2'}{\tau_1 + \tau_2} \left(1 - e^{-\frac{t}{\tau_2'}}\right) \right\} \tag{30}$$

and

TABLE 10. Mobility and Life of Current Carriers in Terphenyl

Carriers	μ_1	μ_2	τ	τ_1	τ_2	τ_2'
	$cm^2/V \cdot sec$		μsec			
Holes	10^{-4}	$2.4 \cdot 10^{-5}$	3	7.5	5	10
Electrons	$1.2 \cdot 10^{-5}$		2.4			

$$
\begin{gathered}
U_{\max}(t, v) = \frac{e \cdot n_0 v}{C_{in} d^2} \mu_1 \tau_2 \left(1 - \frac{\tau_2}{\tau_1 + \tau_2} e^{-\frac{T'}{\tau_2'}}\right), \quad t \to \infty \\
T' = \frac{d^2}{\mu_2 v} - \frac{\mu_1}{\mu_2} \tau,
\end{gathered} \tag{31}
$$

where τ_1 is the capture time of a current carrier by a trap from which release is possible; τ_2 is the lifetime of a current carrier with mobility μ_1 before its capture by a deep trap; τ_2' is the lifetime of a current carrier with mobility μ_2 before its capture by a deep trap; μ_1 is the mobility of a carrier before its capture by any trap; and μ_2 is the mobility of a carrier after its capture by a shallow trap.

Calculation by the equations given yields the values of the magnitudes indicated that are given in Table 10.

As can be seen from the table, the condition $\tau \ll \tau_{tr} \ll \tau_t$ is in fact satisfied, since $\tau_1 = \tau_{tr}$. The mobility may be regarded as the drift mobility, since the acts of trapping the carriers take place repeatedly on their movement through the crystal. An interesting question is that of the nature of the mobility μ_1. Judging from its magnitude it can hardly relate to the microscopic mobility. It is most likely that it is also a drift mobility and is determined by still shallower traps. From this we may conclude that the process of transfer of the charge carriers in polycrystalline samples is far more complex than in monocrystalline samples. In connection with this a fundamental question is how the mechanism of the conduction of polycrystalline samples changes in comparison with monocrystalline samples. Although such comparisons have been made previously in studies of dark conduction, it is important to consider the change in the mobility of the carriers.

Tobin and Sitzer [74] have studied the characteristics of the transfer of photoexcited carriers in a monocrystal and a powdered complex of pyrene and tetracyanoethylene. For monocrystalline samples, an unusually large mobility of the holes at room temperature was obtained: ~ 30 cm^2/V · sec. The mobility of the electrons was three orders of magnitude lower. At the same time, the microscopic mobility of the holes must have been even greater since the measured mobility has an activation dependence on the temperature with an activation energy of 0.5 eV and is apparently the drift mobility. For the powdered samples, the mobility of the holes fell by almost two orders of magnitude, being 0.7 cm^2/V · sec at room temperature.

A still greater change in the mobility of the charge carriers is observed for the liquid state. Measurements of the mobility of holes in crystalline pyrene gave values of 0.35 cm^2/V · sec at 10°C and 0.28 cm^2/V · sec at 155°C. In liquid pyrene, the mobility of the holes falls by almost three orders of magnitude, being 3×10^{-4} cm^2/V · sec at 155°C [75]. This once again shows that the perfection of the structure exerts a decisive influence on the electrical characteristics of low-molecular-weight materials.

Nevertheless, organic semiconductors exist which possess considerable mobilities even in the form of polycrystalline samples. In a study of films of tetracene [61], it was shown that the time of transfer corresponds to a mobility of the holes >0.5 cm^2/V · sec, and at high field strength these holes, after entering the bulk of the specimen, cannot be captured by the traps and all reach the collector electrode. Nevertheless, the charge transferred by them rises with an increase in the attracting field.

A detailed study of this phenomenon enabled Frankevich and Balabanov [61] to develop ideas on the influence of the injection of carriers from the electrode on the transfer process. It was found that an electrode in the immediate vicinity of which holes are generated injects electrons into the film. This explains the following effects: (1) A pulse of the motion of generated holes is observed without the application of an external attracting field thanks to the field of the space charge of the injected electrons. (2) Because of the formation of a space charge, the injected electrons do not pass into the bulk of the sample and the signal from their motion is absent. (3) With an increase in the width of the generation layer the proportion of holes passing into the bulk increases, since the concentration of electrons decreases with increasing distance from the electrode. (4) The dark volt–ampere characteristics have the form typical for a rectifying contact. (5) A photovoltaic effect is observed.

All that has been stated above shows the increased concentration of electrons in the electrode layer. The recombination of the holes with the electrons, which have an increased concentration, competes with the passing of the holes into the bulk of the crystal, which also explains the linear increase in the number of such holes with a rise in the electric field. At small strengths of the extracting field the holes can fall into traps during their travel to the collecting electrode. Since the amplitude of the voltage pulse U(t, v) is determined by the number of moving holes and the distance through which they pass before falling into traps and both these processes are proportional to the extracting field v, then $U(t, v) \propto v^2$. In strong fields, when the holes reach the second electrode without being captured, $U(t, v) \propto v$. The voltage at which the square-root section of this law passes into the linear section is determined by the expression $v = d^2/\mu\tau$, and since $\tau < 5 \times 10^{-7}$ sec the mobility must be of considerable magnitude: ≥ 32 cm^2/V · sec.

Until now, all the results considered could be explained on the basis of the applicability of the band scheme. The small drift mobilities and their activation increase with rising temperature observed in a number of cases are explained by the presence of trapping levels in the specimens and by the repeated capture of the charge carriers in their motion through the sample. However, experimental results exist which are not compatible with the idea of the band model.

The nature of the hole mobility that was studied in monocrystals of ferrocene is not clear [56]. At room temperature, the mobility is 2.2×10^{-2} cm^2/V · sec and it increases with a rise in the temperature (see Fig. 23) with an activation energy close to 0.01 eV. Motion limited by traps presupposes that $\mu_{dr}/\mu \ll 1$. According to equation (19), this condition requires that $N_s/N_{tr} < 1$, since the exponent at such small values of ΔE is close to unity over the whole range of temperatures considered. If the band is narrow, as the motion of carriers with a small mobility requires, N is of the order of the concentration of molecules, and the density of the trapping centers must exceed the density of the molecules. Thus, the mechanism considered is contrary to the actual structure of the crystal.

It may be assumed that in a monocrystal of ferrocene there is no intermolecular jump mechanism of conduction. If the probability of the transfer of a charge from one molecule to the neighboring one is w and the distance between the interacting molecules is a, the diffusion coefficient $D = wa^2$. The mobility in the case of a jump mechanism is determined from Einstein's relation $\mu/D = e/kT$:

$$\mu = \frac{a^2 e}{kT} w. \tag{32}$$

The limit of applicability of the band model can be determined approximately [76]. In order to consider a definite mechanism of scattering described by relaxation time τ the following inequality must be satisfied in accordance with the principle of indeterminacy

$$\tau_r > \frac{h}{\chi},$$

where χ is the width of the conduction band.

If we estimate the width of the band as $\hbar^2 k_0^2/2m$, where k_0 is approximately equal to π/a, we find that

$$\mu = \frac{e}{m}\tau_r > \frac{4}{\pi}\frac{ea^2}{\hbar} \approx 1 \text{ cm}^2/\text{V}\cdot\text{sec} \tag{33}$$

for the usual values of the lattice constant a. Thus, a simple consideration shows that the band mechanism is applicable if the microscopic mobility exceeds unity. At lower mobilities, the jump mechanism of the motion of the charges must be considered.

In the movement of nonequilibrium carriers through a molecular crystal both the drift and the jump mobility may be less than unity in absolute magnitude and increase with a rise in the temperature by the activation law. However, the jump mobility is far less sensitive to structural defects of the type which exert a decisive influence on the characteristics of the drift mobility.

Direct confirmation of the existence of jump mobility can be obtained by studying systems consisting of a matrix neutral with respect to the transfer processes with included molecules of a semiconductor. In this case, by varying the concentration of molecules with conjugated double bonds, it is easy to change the distance between them and to determine the characteristics of the jump mobility of the carriers. Such a study has been carried out on solid solutions of the leuco base of malachite green in various organic media [57]. The main results were obtained using poly-(vinyl chloride) as a polymeric matrix and sucrose benzoate as a vitreous matrix. The carriers were generated by light at one of the electrodes and the drift pulse of the photoexcited carriers through the sample was measured. The organic substances used as matrices were insulators ($\rho > 10^{17}\,\Omega \cdot \mathrm{cm}$) and did not possess photoconductivity. Likewise, the photoconductivity of the matrix was not sensitized by the leuco base of malachite green. Thus, the matrix was neutral with respect to the generation process and with respect to the process of the transfer of the charge carriers in the samples. The mobility of the holes calculated on the basis of the measurements of the transit time proved to be 5×10^{-2} cm^2/V·sec. No signal from the motion of the electrons was observed and, consequently, it may be concluded that the mobility of the electrons was considerably smaller than the mobility of the holes.

In addition to this, the stationary photoconductivity when the specimen was irradiated with visible light was studied. It was found that the mobility and the photocurrent do not depend on the material of the matrix. A photocurrent is observed only at high concentrations of the leuco base. The dependence of the mobility on the temperature and concentration of the leuco base could not be measured, but the activation energy of the photocurrent was 0.1 eV and the photocurrent rose very greatly with an increase in the concentration of the leuco base. The experimental results confirm that in this system a jump mechanism of conduction is in fact present.

The quantitative characteristics of the jump process were obtained in a study of solutions of the free radical diphenylpicrylhydrazyl in benzene [77]. The EPR spectra of concentrated solutions are characterized by a single narrow line, which shows a considerable exchange of unpaired electrons. Starting from this, it may be assumed that unpaired spins moving in the solution from one molecule of diphenylpicrylhydrazyl to another by a jump mechanism take part in conduction. The volt–ampere characteristics of the dark currents were linear in the region of high concentrations (>0.01 mole/liter) and deviated from linearity at low concentrations. In the latter case, the current varied as $i \propto v^c$, where c is an experimental constant less than unity and decreases with a reduction in concentration. In the field of low concentrations, the EPR spectra consist of broad lines with a hyperfine structure of five components and it may be concluded that the unpaired electrons are localized on the neighboring nitrogen atom. Here, apparently, there is a conduction mechanism which differs from the jump mechanism observed in the region of high concentrations. For jump conduction, in accordance with equation (32), it is possible to write

$$\sigma = ne\mu = n\frac{a^2e^2\nu_t}{kT}\exp\left(-\frac{\Delta E}{kT}\right), \tag{34}$$

where n is the concentration of molecules of α,α-diphenyl-β-picrylhydrazyl.

The probability of a jump is $w = \nu_t \exp(-\Delta E/kT)$, where ΔE is the height of the barrier for a jump and ν_t is a frequency factor. The activation energy ΔE obtained from the temperature dependence of the conductivity does not depend on the concentration of the solution and is 0.37 ± 0.02 eV. The exchange of unpaired electrons depends on ν_t but not on ΔE, since the value of the activation energy of the conduction is the same in the regions of high and low concentrations. The nature of the solvent has an influence on ΔE. Consequently, the activation energy in solid benzene is greater than in liquid benzene.

It is interesting to note that the infrared spectra of solutions of diphenylpicrylhydrazyl in benzene, which are quite similar to the spectra of pure benzene, exhibit an absorption band at 0.37 eV which coincides with the activation energy of conduction. For the electronic jumps, ν_t is determined by the relation $\nu_t = I/2h$, where I is the exchange energy, which is calculated from the EPR spectra. In the general case, I depends on the ratio of the jump distance a to the effective radius of an unpaired electron r and is written as $I = I_0 \exp(-a/r)$. If one also takes into account the fact that the exchange energy depends on the angle between the axes of symmetry of neighboring molecules of α,α-diphenyl-β-picrylhydrazyl, an anisotropy factor A_0 must be introduced into the formula for electrical conductivity (34) and in the final form it becomes

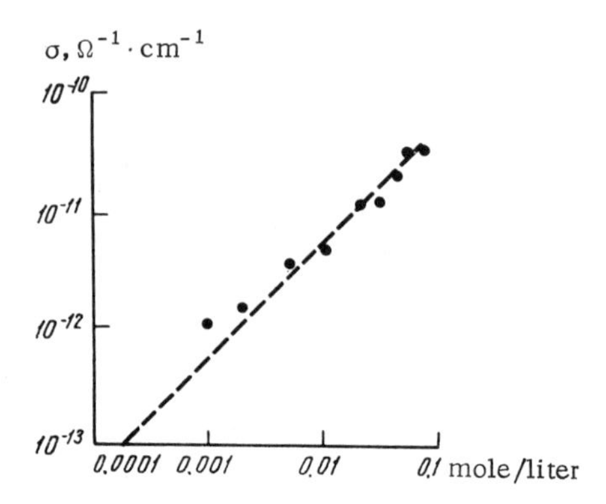

Fig. 25. Conductivity of solutions of α,α-diphenyl-β-picrylhydrazyl in benzene as a function of the concentration. Broken line—calculated from equation (35).

$$\sigma = A_0 n \cdot \frac{a^2 e^2 \cdot I}{2hkT} \exp\left(-\frac{\Delta E}{kT}\right). \tag{35}$$

This formula expresses the concentration dependence of the conductivity in the region of high concentrations. In Fig. 25 the results of experimental measurements of σ and the figures calculated from formula (35) are compared. Good agreement is observed between the calculated and experimental values of the conductivity at high concentrations and a deviation at concentrations of less than 10^{-2} mole/liter. The results convincingly favor the jump mechanism of conduction in solutions of an organic semiconductor—α,α-diphenyl-β-picrylhydrazyl.

Analogous effects should be observed for other compounds of this class, as well.

MOBILITY OF CURRENT CARRIERS IN POLYMERIC SEMICONDUCTORS

In crystalline low-molecular-weight organic semiconductors, slight structural changes or the presence of small amounts of other substances greatly complicate the transfer mechanism and do not enable the drift mobility to be connected quantitatively with the defects. It is probable that for polymeric semiconductors the transfer mechanism is still more complicated.

An investigation of the characteristics of the current carriers in an electric field of low-energy electrons injected by a pulse has been carried out in films of semiconducting polymers obtained by the radiation-thermal treatment of saturated linear polymers [62, 63, 78]. The results obtained characterize the movement of the holes. The signal from the electrons was detected clearly in relatively highly conducting samples, but it was too small for the performance of quantitative measurements. With an increase in the strength of the electric field, T_t, determined by the time to achieve the maximum amplitude of the voltage pulse, decreased, which shows that the holes reached the collector electrode. Figure 26a shows the results of measurements at various temperatures. In all the cases considered above, the relationship between $1/T_t$ and E was linear. According to equation (17) this means that the mobility does not depend on the field strength. The distinguishing characteristic of polymeric semiconductors consists in the fact that the relationship between $1/T_t$ and E is nonlinear, i.e., the mobility increases with a rise in the electric field.

It is interesting that, at the field strengths studied, in low-molecular-weight organic semiconductors the mobility does not depend on the field strength [60]. The static volt–ampere characteristics were read simultaneously (Fig. 26b). The relationship between i and E has a linear section in weak fields $< 10^3$ V/cm; in strong fields the current rises exponentially with an increase in E, but this is not connected with the evolution of Joule heat in the film, which cannot be neglected if it is greater than 0.1 W/cm^2. As was confirmed by special experiments, the power dispersed in the samples was considerably less than this value.

The results given provide the possibility of unambiguously connecting the conductivity of films of polymeric semiconductors with the mobilities of the charge carriers. Since the current of the holes generated in the electrode layer is equal to within an order of magnitude to the dark current, it may be considered that the motion of the nonequilibrium and equilibrium charge carriers takes place under similar conditions and the measured mobility is the ohmic mobility and relates directly to the dark conductivity. In its dependence on the temperature and the outside field, the conductivity is expressed in the general form as

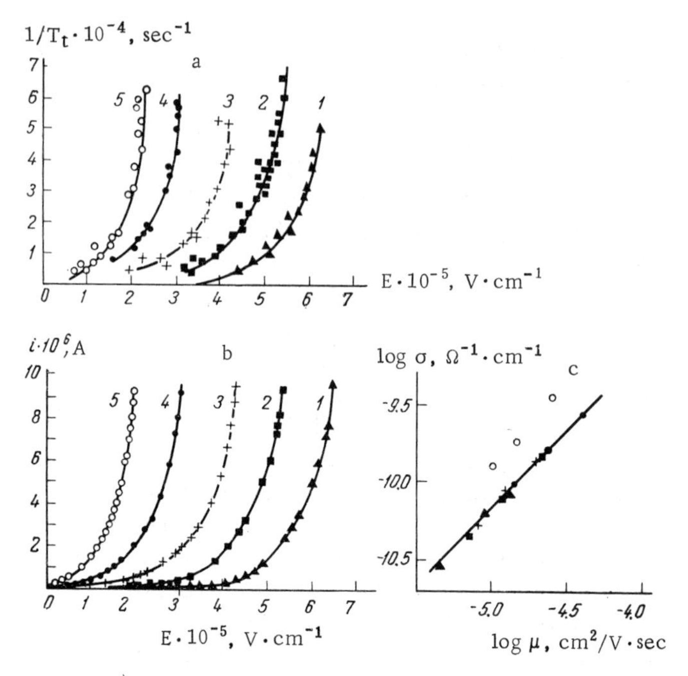

Fig. 26. $1/T_t$ as a function of E (a), i as a function of E (b), and σ as a function of μ (c) for modified poly(vinyl acetate). 1) −11°C; 2) +18°C; 3) +40°C; 4) +72°C; 5) +110°C.

$$\sigma(E, t^\circ) = ep(E, t^\circ)\mu(E, t^\circ). \tag{36}$$

If the concentration of holes $p(E, t^0)$ does not depend on the field and the temperature, $\log \sigma (E, t^0) = \text{const} + \log \mu (E, t^0)$ and on a double logarithmic scale the values of σ and μ for the same values of E and t^0 must lie on a single straight line at an angle of 45° to the axes of coordinates. In actual fact, as can be seen from Fig. 26c, the values of σ and μ calculated from independent measurements of i and T for three arbitrary values of E at each temperature lie on a single straight line and only at 110°C do the points not coincide with the straight line. Apparently, in this region of temperatures the thermal liberation of carriers begins to have an effect. It is true that here, also, the concentration of carriers does not change with a change in the applied field. This follows from the constancy of the angle of inclination of the relation log σ versus log μ. The concentration of holes is found to be 7 ×

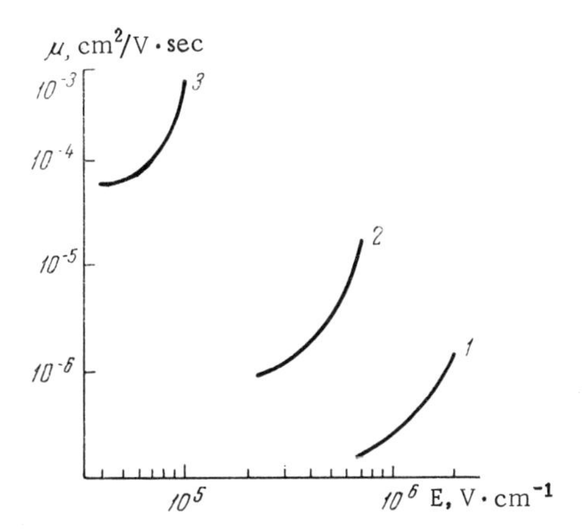

Fig. 27. Mobility of holes at room temperature for samples of modified polyethylene obtained at 290°C (1), 450°C (2), and 570°C (3).

10^{13} cm^{-3}. These results show that the change in σ with a change in the temperature and in the applied field at sufficiently low temperatures is determined completely by the dependence of the mobility on these parameters. The activation energy of conduction within the limits of experimental accuracy coincides with the activation energy of mobility, since

$$\Delta\sigma(t^{\circ}) = \text{const}\ \Delta\mu(t^{\circ}). \tag{37}$$

With an increase in the temperature of heating of the samples, the dark currents (conductivity) rise for one and the same values of E at room temperature, the mobility rising simultaneously. Figure 27 gives the mobility at room temperature for samples with temperatures of heating of 290, 450, and 570°C calculated from the experimentally obtained values of the transit time. The minimum values of T_t measured in films of semiconducting polymers are $\mu \approx 10^{-3}$ $cm^2/V \cdot sec$. With a further increase in the temperature of heating, the dark currents become so large that the pulse method becomes inapplicable and it can only be assumed that the mobility increases even further with an increase in conductivity.

CONDUCTION MECHANISM IN MOLECULAR CRYSTALS

A calculation of the band structure of organic semiconductors is very important for an explanation of the mechanism of conduction. As shown above, for the band approximation to be applicable several conditions imposed by the principle of indeterminacy must be satisfied. In Fröhlich and Sewell's formulation [79] this leads to the situation that in the band approximation for substances with a mobility of the order of 1 $cm^2/V \cdot sec$, as is the case for anthracene, the width of the band is of the order of kT or less at room temperature, and the thermal velocity, the maximum value of which is determined by the expression $v_{max} \approx \chi a/\hbar$ (where χ is the width of the zone), in this case is 10^5 cm/sec to an order of magnitude.

The calculation of the band structure of anthracene has been carried out by Le Blanc [69]. A crystal of anthracene forms a monoclinic lattice with two molecules per elementary cell. If the vectors of the elementary cells are denoted by $\bar{\mathbf{a}}$, $\bar{\mathbf{b}}$, and $\bar{\mathbf{c}}$, and the new vectors $\bar{\alpha} = 1/2\ (\bar{\mathbf{a}} + \bar{\mathbf{b}})$ and $\bar{\beta} = 1/2\ (-\bar{\mathbf{a}} + \bar{\mathbf{b}})$ are introduced, then $\bar{\alpha}$, $\bar{\beta}$, and $\bar{\mathbf{c}}$ are the geometrical centers of neighboring molecules. Such a replacement enables us to obtain intrinsic values in each unambiguous function **k**.

Since the energy of interaction between molecules is small in comparison with the distance between the energy levels of isolated molecules, the approximation of the strong bond can be used. The wave function of a superfluous electron can be expressed by the Bloch sum of the molecular orbitals:

$$\Psi_k(\bar{\mathbf{r}}) = \sum_{n=1}^{N} \exp(i\overline{\mathbf{k}\mathbf{r}_t})\, \varphi_n(\bar{\mathbf{r}} - \bar{\mathbf{r}}_t). \tag{38}$$

Here $\bar{\mathbf{r}}_t$ determines the geometrical center of the molecule n. The superposition of the periodic boundary conditions determines the elementary cell in k-space, the volume of which is given by the conditions

$$-\pi < \overline{\mathbf{k}\alpha},\ \overline{\mathbf{k}\beta},\ \overline{\mathbf{k}\mathbf{c}} \leqslant +\pi. \tag{39}$$

Condition (39) does not define the usual Brillouin zone but, since it has correspondence with the points in the Brillouin zone and the points in the elementary cell determined by equation (39) and the latter is far simpler to consider, it is more convenient to use equation (39) to determine the limits of **k**.

In the case of an electronic zone, the wave functions φ_n must describe the state when a superfluous electron is localized on the molecule n. Consequently, φ_n is the wave function of the electron in the ground state of a singly charged negative ion. The interaction of the superfluous electron with the other electrons and charges in the crystal can be expressed by Hartree's equation. The crystal field is expressed as

$$V(\bar{\mathbf{r}}) = \sum_n V_n(\bar{\mathbf{r}} - \bar{\mathbf{r}}_n), \tag{40}$$

where V_n is the Hartree potential of an isolated neutral molecule. Then the intrinsic value of ψ_n is given by

$$E(\bar{\mathbf{k}}) = E_0 + E_1 + 2\sum_s E_s \cos \bar{\mathbf{k}}\bar{\mathbf{r}}_s, \tag{41}$$

where

$$E_0 = \int \varphi_n [-(\hbar^2 \Delta/2m) - eV_n] \varphi_n \, d\tau;$$

$$E_1 = \sum_s \int |\varphi_n|^2 V_{n+s} \, d\tau; \quad s \neq 0;$$

$$E_s = \int \varphi_{n+s} V_{n+s} \varphi_n \, d\tau.$$

The band structure is determined by the intermolecular resonance integrals E_s. To calculate them we can take the potential of a neutral molecule V_n in the following form [80]:

$$V_n(\bar{\mathbf{r}} - \bar{\mathbf{r}}_n) = -\sum_A \left(\frac{Ze^2}{R_A}\right) + 2\sum_{i=0}^{a} I_n^i, \tag{42}$$

where R_A is the distance from the A-th nucleus of the n-th molecule and I_n^i is the Coulomb potential of the i-th molecular orbital molecule. We can take φ_n as the linear combination of atomic 2p orbitals, using for the electron and hole bands the corresponding Hückel coefficients for the lowest nonbonding and highest bonding π-orbitals. Here it is possible to consider only two-center integrals, since the contribution from multicenter integrals is negli-

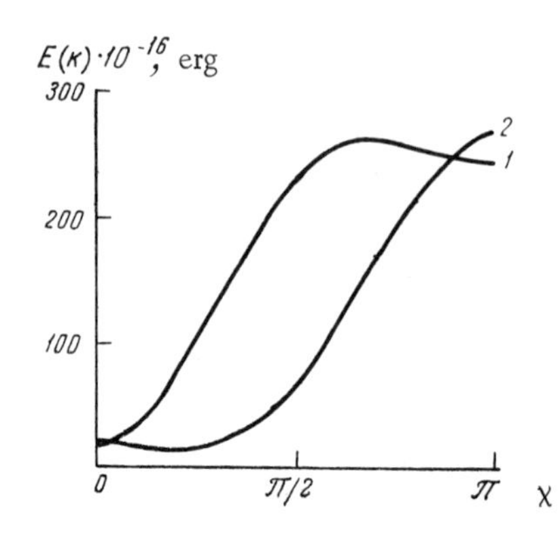

Fig. 28. $E(\bar{\mathbf{k}})$ as a function of χ for anthracene. 1) Hole; 2) electron.

gible. The resonance integrals rapidly decrease with the intermolecular distance and, as has been found [69], only the interactions with the five nearest neighboring molecules are important. The integrals between the molecules connecting $\bar{\alpha}$ and $\bar{\beta}$ are equal and can be written as E_α. Other important members are $E_b(\bar{\mathbf{r}}_s = \bar{\mathbf{b}})$ and $E_\gamma(\bar{\mathbf{r}}_s = \bar{\mathbf{c}} + \bar{\alpha}$ and $\bar{\mathbf{r}}_s = \bar{\mathbf{c}} - \bar{\beta})$.

In this case, the energy in the one-electron approximation for each band, excluding the constants E_0 and E_1, defined by equation (41) can be written

$$E(\bar{\mathbf{k}}) = 2E_\alpha[\cos\bar{\mathbf{k}}\bar{\alpha} + \cos\bar{\mathbf{k}}\bar{\beta}] + E_b \cos\bar{\mathbf{k}}(\bar{\alpha} + \beta) + 2E_\gamma[\cos\bar{\mathbf{k}}(\bar{\mathbf{c}} + \bar{\alpha}) + \cos\bar{\mathbf{k}}(\bar{\mathbf{c}} - \bar{\beta})]. \quad (43)$$

A maximum and a minimum in $E(\bar{\mathbf{k}})$ are observed when $\bar{\mathbf{k}}\cdot\bar{\mathbf{c}} = 0$ and $\overline{\mathbf{k}\alpha} = \overline{\mathbf{k}\beta} = \chi$, i.e., for $\bar{\mathbf{k}}$ parallel to $\bar{\mathbf{b}}$. The shape of $E(\bar{\mathbf{k}})$ for each zone along this direction $(\bar{\mathbf{k}})$ is shown in Fig. 28. As follows from the figure, the width of each band is approximately 2.4×10^{-14} erg or 0.56 kT at room temperature. In a further analysis of the structure of the bands, the problem was solved by considering the wave functions of the whole crystal and not of a superfluous electron or hole. More accurate calculations [72] taking into account the effects of intermolecular electronic interaction and molecular vibration lead to similar results. The widths of the conduction band and the valence band are 0.02 eV or less, in good agreement with the values obtained previously. If the scattering of the charges can be described by the introduction of the relaxation time $\tau_r(\bar{\mathbf{k}})$, for the velocity of the carriers the following equation is satisfied:

$$v(\bar{k}) = \frac{1}{\hbar} \frac{\partial E(\bar{k})}{\partial \bar{k}}. \qquad (44)$$

In considering the approximation of a constant relaxation time independent of the energy ($\tau_r(\bar{k}) = \tau_r$), the following formula is satisfied for the components of the mobility tensor

$$\mu_{ij} = e\tau_r < v_i v_j > \frac{1}{kT}, \qquad (45)$$

where v represents the corresponding components $v(\bar{k})$ and the brackets indicate the statistical mean over the whole zone. Since the width of the zone is ~ 0.56 kT, the Boltzmann probability changes only by a factor of 1.7 within the limits of the whole zone $\langle v_i v_j \rangle$ and therefore does not in fact depend on the temperature. By neglecting the change in the Boltzmann factor it is possible to determine $\langle v_i v_j \rangle$ and calculate the mean square velocity as $\langle v^2 \rangle = \langle v_a^2 \rangle + \langle v_b^2 \rangle + \langle v_c^2 \rangle$. The value of the velocity is estimated as relatively small ($\sim 10^5$ cm/sec) in agreement with the small widths of the bands.

Many important properties of inorganic semiconductors are determined by the high mobilities of the charge carriers. In order to determine the potential possibilities of using organic compounds with conjugated bonds just as semiconductors it is necessary to investigate what characteristics of the mobility of the carriers can be achieved in them. Even from a consideration of the experimental data on the measurement of the mobility the preliminary conclusion was drawn that in organic semiconductors the movement of charges is possible both by the band and by the jump mechanism. The two mechanisms must appear in the different dependences of the mobility on such factors as the temperature, the anisotropy of the lattice, and the magnitude of the mobility. For organic semiconductors a comparison of the two mechanisms of mobility has been carried out by Glarum [81] taking into account the known structure of the bands in organic molecular crystals.

Bearing in mind the small degree of overlapping of the wave functions in a molecular crystal and the narrow bands of the order of kT, as calculated for anthracene, let us consider the approximation of the force bond and calculate the relaxation time for the delocalized or band mechanism and the localized or jump mechanism.

In the consideration of the band mechanism, the mobility measured in the direction parallel to an electric field applied along the x axis of the crystal is expressed as

$$\mu = \frac{e}{kT} \langle v_x^2 \tau \rangle \tag{46}$$

provided that the density of the carriers is sufficiently low and Boltzmann statistics are applicable. The brackets indicate the mean value of the magnitudes in the reciprocal lattice space, the Boltzmann exponential member being taken as a weighted factor, and v_x is the x-component of the velocity of the carriers.

Where the width of the zone is far smaller than kT, the energy levels in the band are also expressed by

$$\mu = \frac{ea^2}{\hbar} \frac{1}{6\pi\gamma} \frac{\chi^3}{\hbar\omega_0 (kT)^2} \tag{47}$$

where a is the lattice constant, 2χ is the width of the zone, ω_0 is the frequency of vibrations of the lattice, and γ is a dimensionless parameter of the electron-lattice interaction which is different in the two models considered. The factor $ea^2/\hbar$ has the dimensions of mobility and for ordinary intermolecular distances amounts to 1-2 $cm^2/V \cdot sec$.

If the width of the band is far greater than kT, the usual expression for mobility is obtained:

$$\mu = \frac{ea^2}{\hbar} \frac{1}{4\pi\gamma} \frac{\chi}{h\omega_0} \left(\frac{2\pi\chi}{kT} \right)^{3/2}, \tag{48}$$

since χ is inversely proportional to the effective mass and γ is proportional to the square of the deformation potential. Qualitatively, the anisotropy effect is taken into account by replacing the member $(2\pi\chi/kT)^{3/2}$ in equation (48) by the factor $(2\pi\chi/kT)^{n/2}$.

For the band model to be correct, two conditions must be satisfied:

$$\chi \gg \hbar\omega_0, \qquad \chi\tau \gg \hbar,$$

and in the case of narrow bands

$$\mu \gg \frac{ea^2}{\hbar} \frac{\hbar\omega_0}{kT}. \tag{49}$$

Since the energies $\hbar\omega_0$ and kT are usually of the same order of magnitude, it follows from condition (49) that the band mechanism is applicable only where the mobility is greater than ~1 $cm^2/V \cdot sec$.

For the jump mechanism, the mobility is expressed by equation (32):

$$\mu = \frac{ea^2}{kT} w.$$

The probability of a transition in unit time is

$$w = \frac{2\pi\beta^2}{\hbar^2\omega_0} \exp[-\gamma(2S+1)] I_0 \{2\gamma [S(S+1)]^{\frac{1}{2}}\}, \quad (50)$$

where $S = \{\exp(\hbar\omega_0/kT) - 1\}^{-1}$, β is the intermolecular resonance energy, and I_0 is a Bessel function.

For the jump mobility, from (32) and (50) we can deduce that

$$\mu = \frac{ea^2}{\hbar} \frac{2\pi\beta^2}{kT\hbar\omega_0} \exp[-\gamma(2S+1)] I_0 \{2\gamma [S(S+1)]^{1/2}\}. \quad (51)$$

In the particular case where $\gamma S = \gamma \frac{kT}{\hbar\omega_0} \gg 1$, the expression actually becomes

$$\mu = \frac{ea}{\hbar} \left(\frac{\pi}{\gamma}\right)^{1/2} \frac{\beta^2}{(\hbar\omega_0)^{1/2} (kT)^{3/2}} \exp\left(-\frac{\gamma\hbar\omega_0}{4kT}\right). \quad (52)$$

The condition for the applicability of this model is $\omega_0 \tau \gg 1$, or

$$\mu \ll \frac{ea^2}{\hbar} \frac{\hbar\omega_0}{kT}. \quad (53)$$

This result satisfies equation (49) and shows that the jump model is applicable only where the mobility is less than 1 $cm^2/V \cdot sec$.

The conclusion obtained is in good agreement with that found previously in a consideration of the experimental results.

Any considerations of the characteristics of the mobility of organic semiconductors are based on the calculation of the electron-lattice interaction parameter γ. Moreover, γ is directly connected with the activation energy of mobility — the parameter usually most frequently obtained in experiment. In the band approximation,

$$\gamma \approx \left(\frac{\chi}{\hbar\omega_0}\right)^2 \cdot \left(\frac{x^2}{\lambda^2}\right), \tag{54}$$

where x is the amplitude of the zero point of the molecular vibrations and λ is a parameter characterizing the decay of the molecular wave function. Typical values of x and λ are 0.1 and 0.2Å, respectively. Since in the band scheme $\chi \gg \hbar\omega_0$, γ should be of the order of unity or higher. In the jump model, γ depends on the nature of the electron–lattice interaction. For polar crystals, estimates of γ lead to values $\gtrsim 10$. In nonpolar lattices the interaction is considerably less. If in this case the main effect is the displacement of only those molecules which immediately surround the section of localized charge, then

$$\gamma \approx \Sigma \frac{(\Delta r)^2}{x^2}, \tag{55}$$

where Δr is the displacement of a molecule from its equilibrium position. Assuming that Δr is several tenths of an angstrom, again we obtain $\gamma \gtrsim 1$.

In addition to the mobility, the temperature dependence of the mobility may be regarded as an important criterion for distinguishing the band and jump mechanism. In the band mechanism, $\mu \sim T^{-n}$, where $n < 2$ if the scattering takes place on acoustic phonons. The accurate value of n depends on the anisotropy and widths of the zone.

For the jump mechanism, the temperature dependence of the mobility is more complex. With strong electron–lattice interaction the mobility rapidly increases with a rise in the temperature. If, however, $\gamma < 10$, the mobility may not change or may decrease with a rise in the temperature. Thus, a positive temperature coefficient of the mobility is not a criterion of the jump mechanism of conduction, except for polar crystals, where γ is large. If the conditions under which equation (52) is true are satisfied, the activation energy of the jumps is of the order of 0.1 eV.

An activation dependence of the mobility on the temperature must apparently be observed in charge-transfer complexes. Numerous estimates, including some from the Hall effect, lead to

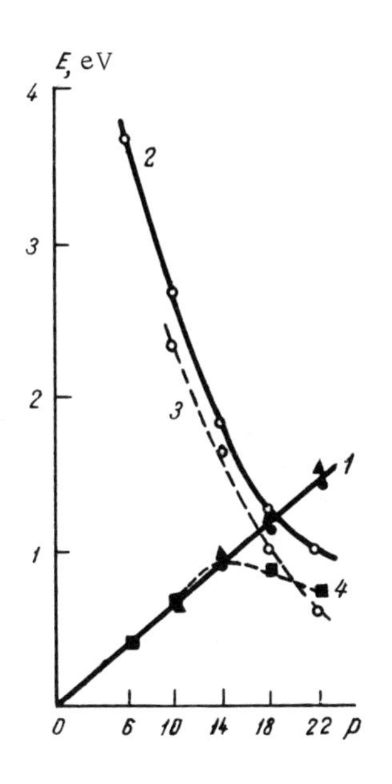

Fig. 29. Dependence of the activation energy of conduction on the number of π-electrons in the molecule of a hydrocarbon. 1) 1/4 resonance energy (●) and the energy calculated for E = 0.067 p (▲); 2) $^3E_1$ (solution) = E_t; 3) $^3E_1$ (solid); 4) experimental values of the activation energy; p = number of π-electrons.

values of $\mu \ll 1$ cm²/V·sec, so that the jump model is applicable. For such low mobilities, equation (51) predicts either a small value of the intermolecular resonance energy ($\beta \ll kT$) or $\gamma S \gg 1$. Since the stability of the complexes is probably connected with a large value of β, the first condition is not satisfied and it may be concluded that the mobility must increase rapidly with a rise in the temperature.

It may be concluded that the mobility is a more reliable criterion for distinguishing the two mechanisms considered. In anthracene, as has been shown, the mobility is ~1 cm²/V·sec and is thus in the transition region. This is confirmed by the fact that an evaluation of γ from equation (47) using the values of the width of the band calculated by Le Blanc (~0.015 ev) and $\hbar\omega_0 \approx 0.01$ eV leads to values of $\gamma \approx 10^{-2}$, which are at least an order of magnitude lower than estimated above. A possible explanation is that anthracene, in accordance with its measured mobility, occupies an intermediate position between the two conduction mechanisms.

This is confirmed by comparing the properties of a sequence of aromatic hydrocarbons. The activation energy of conduction can be compared with the position of the electron levels in the crystal [82]. Figure 29 gives the values of the activation energy of conduction ΔE, the energy of excitation into the triplet state $^3E_1$, and also the energy E_r, proportional to the resonance energy, of benzene, naphthalene, anthracene, tetracene, and pentacene. As can be seen, for the first three members of the series the value of ΔE coincides with the value of E and for the other two hydrocarbons the value of ΔE is close to that of $^3E_1$. It may be assumed that when the lowest excited state $^3E_1$ exceeds the experimentally found value of ΔE, conduction takes place by a jump mechanism directly connected with the resonance energy. When $^3E_1$ becomes less than E, the activation energy of conduction is determined by the thermal filling of the $^3E_1$ level and a band mechanism is observed. The passage of ΔE through a maximum with an increase in the number of π-electrons p in the molecule, as shown in Fig. 29, can be explained from this point of view.

If it is assumed that the current carriers are electrons injected from the electrode and localization takes place in a molecule of hydrocarbon with the formation of a negative ion, a jump takes place between an ion and a neutral molecule. In a general case, the larger the polyconjugated molecule the stronger the reaction with an electron because of the increase in resonance energy. It follows from expression (52) that the greater the parameter of electron–lattice interaction γ the greater is the activation energy of jump mobility. Taking into account the fact that the degree of resonance stabilization of a negative ion is proportional to γ, it may be expected that for the jump mechanism the activation energy will grow with an increase in the number of π-electrons in the molecule, as is found experimentally. An empirical relation between p and E_r gives the expression $E_r = 0.067$ p and, in this case E_r is proportional to one quarter of the resonance energy.

With a further increase in the dimensions of the polyconjugated molecules a transition takes place to the band mechanism and the activation energy of the generation of carriers through excited triplet states falls. The formation of the carriers apparently takes place by the interaction of two excitons. As can be seen from the figure, anthracene is in fact close to the transition region between the two conduction mechanisms, although it always remains in the region of jump mobility. Recent calculations show that for anthracene the jump mechanism is probably preferable [83]. At the same time, the simultaneous measurement of the drift and Hall mobilities have shown that $\mu_{dr} = 1.4\ cm^2/V \cdot sec$ and $\mu_H = 35$

$cm^2/V \cdot sec$, i.e., $\mu_H/\mu_{dr} = 25$. The theory predicts a negative sign of the Hall effect in anthracene, while the width of the band for these values of the Hall mobility is evaluated as 4×10^{-3} kT [84]. Thus, the question of whether the band mechanism is correct still remains open for anthracene. The band mechanism is observed in tetracene. This agrees with the conditions necessary for the appearance of the band mechanism given above, since the mobility in tetracene is greater than 1 $cm^2/V \cdot sec$ [61].

The establishment of the basic characteristics of the band structure makes it possible to consider the kinetic properties of organic semiconductors observed on the application of electric and magnetic fields, a temperature gradient, and so on. The calculated kinetic coefficients of organic semiconductors have been given by Kubarev and Mikhailov [85, 86]. On considering a cubic lattice, the potential of the interaction of an electron with the lattice vibrations can be taken in the form

$$V = E_1 \operatorname{div} \bar{\mathbf{u}}, \tag{56}$$

where E_1 is the constant of the deformation potential and $\bar{\mathbf{u}}$ is the vector of the displacement of the lattice. The wave functions of an electron in the approximation of the force bond consist of a linear combination of one-electron molecular wave functions φ_n [cf. equation (38)] and the matrix element of the energy of perturbation is found (56). This makes it possible to calculate the probability of the transition $\bar{\mathbf{k}} \rightarrow \bar{\mathbf{k}}'$ for an electron in 1 sec and to formulate the usual kinetic equation. In the solution of the kinetic equation, it is assumed that the partition function of the phonons remains an equilibrium function, and the partition function of the electrons differs little from the equilibrium function, which does not depend on the wave vector $\bar{\mathbf{k}}$. The latter is justified for semiconductors with a narrow conduction band, when the following condition is satisfied.

$$\chi \ll \Delta E = F, \tag{57}$$

where F is the distance from the Fermi level to the conduction band. In organic semiconducting crystals the inequality (57) is satisfied, since $\chi \lesssim kT$ [69, 72]. It is also assumed that the electrons interact only with longitudinal phonons. Under these assumptions, the relaxation time is expressed as

$$\tau(\bar{\mathbf{k}}) = \frac{N}{\Omega} \frac{\pi\hbar M v^2 \chi a}{2E_1^2 kT} \cdot \frac{1}{D_1(\bar{\mathbf{k}}) + D_2(\bar{\mathbf{k}}) - \frac{1}{2} D_3(\bar{\mathbf{k}})}, \tag{58}$$

where Ω is the volume of the crystal, N is the number of molecules, M is the mass of a molecule, and d_1, d_2, and d_3 characterize the electron transitions with the absorption and emission of phonons. The relaxation time is a slowly changing function of $(\mathbf{k})$ and therefore it may be assumed that

$$\tau(\bar{\mathbf{k}}) = \tau_r = \frac{N}{\Omega} \frac{\pi\hbar M v^2 \chi a}{2E_1^2 kT} \beta^2, \tag{59}$$

where

$$\beta = \frac{2\hbar v}{\chi}.$$

Using the equations obtained for the relaxation time it is possible to calculate the kinetic coefficients by means of the usual formulas. Thus, the electrical conductivity is expressed by the formula

$$\sigma = \frac{N}{\Omega} \frac{e^2 \chi^3 M v^2 a^2}{64\pi\hbar E_1^2 (kT)^2} \exp\left(-\frac{F}{kT}\right) = \sigma_0 \exp\left(-\frac{F}{kT}\right). \tag{60}$$

The magnitude of σ can be determined by using the values of the constants that are characteristic for organic semiconductors:

$$\frac{\chi}{kT} \sim 1,\ a \sim 10^{-7} - 10^{-8}\ \text{cm},\ M \sim 10^{-22}\ \text{g},\ v \sim 10^5\ \text{cm/sec}$$

In this case,

$$\sigma_0 \approx 100 \left(\frac{\chi}{E_1}\right)^2 \Omega^{-1} \cdot \text{cm}^{-1}. \tag{61}$$

The mobility may be obtained from $\mu = \sigma/ne$, taking into account the fact that the concentration of carriers in the conduction zone

$$n = \frac{N}{\Omega} \exp\left(-\frac{F}{kT}\right),$$

$$\mu \approx 10^{-2} \left(\frac{\beta}{E_1}\right)^2 \text{cm}^2/\text{V} \cdot \text{sec}. \tag{62}$$

The coefficient of the thermo-emf is given by the simple formula

$$\alpha = \frac{\Delta E}{eT}. \tag{63}$$

It is also possible to give an equation for the magnetic resistance and the Hall constant. The Hall constant is defined by

$$R = \pm \frac{6.75 \cdot 10^{17}}{n} \cdot \frac{kT}{\chi} \quad \text{cm}^3/\text{C}. \tag{64}$$

This expression was obtained on the assumption of a weak field the upper limit of which can be evaluated from the inequality

$$H \ll \frac{2c\hbar^2}{e\chi^2\tau_r a^2}. \tag{65}$$

By substituting in this expression constants characteristic for organic semiconductors, we find $H \ll 10^7$ Oe. This value of the strength of the magnetic field, which is large in comparison with the values customary for inorganic semiconductors, is related to the large effective mass and short relaxation time in organic semiconductors. It is interesting to compare the value of the Hall constant obtained from equation (64) with that observed experimentally. This can be done for phthalocyanine, on which the most accurate measurements have been carried out [9, 11].

According to formula (64), $R = 10^{11}$ cm³/C with the substitution of n calculated using the values of F obtained from measurements of electrical conductivity. Experimentally, it was found that $R = 10^{12}$ cm³/C.

The results given were obtained from a consideration of a cubic lattice. The majority of organic semiconductors satisfying the conditions given possess a monoclinic lattice. If one considers the interaction of electrons only with acoustic phonons, for a monoclinic lattice formulas analogous to those considered above are obtained.

In the calculation of the kinetic coefficients the question arises of the limits of application of the kinetic equation. The conditions of applicability of the kinetic equation can be written in the form

$$\tau_r \gg \frac{\hbar}{kT}. \tag{66}$$

Taking into account the fact that kT $\sim \chi$ and replacing kT by χ once, from equation (59) we find the following criterion:

$$\frac{E_1}{\chi} \ll \sqrt{\frac{N}{\Omega} \frac{\pi\hbar M v^2 \chi \beta^2 a}{kT}} . \tag{67}$$

If in (67) we substitute the constants characteristic for organic semiconductors, this equation passes into the condition $\chi \gg E_1$. From this criterion it is more convenient to pass to mobilities. Strictly speaking, the question of the limit of applicability of the kinetic equation can be solved only when the magnitude of the constant of the deformation potential E_1 is known. However, if the relation $\chi / E_1 \sim 10$ is assumed to be reasonable, it follows from equation (62) that the kinetic equation is applicable if $\mu \gtrsim 1$ cm^2/V · sec. This condition shows once more that the general ideas developed for inorganic semiconductors can be applied to organic semiconductors with mobilities greater than 1 cm^2/V · sec. This is understandable, since the band model and the kinetic equation require the introduction of the relaxation time.

CONDUCTION MECHANISM IN POLYMERIC SEMICONDUCTORS

A study of the mechanism of conduction in polymeric organic semiconductors is a considerably more complex problem than in the case of low-molecular-weight crystals and is connected above all with the indeterminacy of the structure of the substances studied. In considering polymeric semiconductors very diverse assumptions frequently having only an extremely remote connection with the actual picture are made. Supplementing this, the number of methods that can be used to establish the electrical characteristics in polymers is frequently still smaller than in the study of low-molecular-weight crystals. Thus, even the method of measuring the drift mobility of injected carriers, not to speak of the Hall effect, can prove ineffective.

In some papers, the tunnel mechanism of conduction is proposed, bearing in mind the fact that the energy spectrum in a macromolecule has a quasimetallic nature [87-90]. For protein molecules, a model has been proposed which consists of a combination of the band model and the tunnel mechanism [91-94].

In many investigations it is assumed that the polymers studied are intrinsic semiconductors [24-26, 95-99]. From electrical measurements, the values of the concentration of carriers and the mobility are calculated from the formulas of the band theory. In spite of the different points of view, certain fundamental conclusions can be drawn from a large number of investigations [16-23, 32, 38, 39].

In the first place it must be mentioned that for polymeric semiconductors the mechanism of conduction is based on the model of the structure of the polymer. The structure of organic polymeric semiconductors deduced from electrical measurements has been studied for various materials with the aid of frequency methods of investigating the dependence of conductivity on the field strength [32, 38] and from current noise [48]. As already mentioned (p. 68), all these methods show that it is necessary to isolate two stages in the movement of a current carrier — motion within the macromolecule and passage from one macromolecule to another.

It is interesting that even the dimensions of the regions of polyconjugation are found to be of approximately the same order of magnitude in all cases: from several tens to several hundreds of angstroms. There is no single point of view concerning the generation and movement of carriers in the macromolecule itself. The region of polyconjugation may be considered as a region having metallic conductivity or as a semiconductor for which the band model is valid [100]. In the first case, the whole of the observed activation energy of conduction is connected with a transition from one region to another. The second case is more complex. If the region of polyconjugation is an intrinsic semiconductor, the activation energy is composed of two parts — the activation energy of conduction in the macromolecule and the activation energy of the jumps. If the region of polyconjugation is a semiconductor with an impurity, the activation energy within the macromolecule does not coincide with the forbidden gap and may even be close to zero as in the case of a degenerate semiconductor; in this case the mechanism must resemble that proposed for doped germanium and silicon at low temperatures [36, 37].

Experimentally, of course, it is always the total magnitude of the electrical conductivity and the activation energy that are measured and therefore it is least of all possible to draw a conclusion on the conductivity of the macromolecules themselves. On the contrary, in theory the most accessible for solution is the problem in which the individual molecule is considered. If the polymeric molecule is considered as a one-dimensional conductor [101] with bands of permitted energies in which there are quasi-free electrons forming a degenerate gas, then in its movement through the molecule the electron can be scattered only in the reverse direction.

Since the number of phonons with the quasimomentum necessary in this case is small, the resistance to the motion of the electron arising with scattering on thermal vibrations is negligibly small. In the case of a low concentration of free electrons conduction has the normal characteristics. If "impurities" are present in the molecule — side groups, disturbances of the chain of conjugation, and so on — the conductivity does not depend on the temperature.

The semiconductor spectrum for a conjugated molecule can be obtained in various ways — for example, by means of the so-called free electron model, in which the macromolecule is considered as a square potential well with infinitely high walls, or by different variants of the method of molecular orbitals [102, 103].

Information on the properties of macromolecules can be obtained at the present time only indirectly from a combination of the methods of measurement at different frequencies, the thermo-emf, and the method of injecting carriers. It is found that the conductivity in regions of conjugation exceeds the overall value by several orders of magnitude [40].

Attempts can also be made to evaluate mobility in the regions of polyconjugation themselves. For this purpose we must know the concentration of electrons in the macromolecules. Several methods are known by means of which it is possible to determine these magnitudes. Measurement of the thermo-emf in thermally treated polyacrylonitrile gives a concentration of electrons of between 10^{13} and 10^{19} cm^{-3}, depending on the thermal treatment. The same results are obtained in a study of modified polyethylene by the EPR method and by measuring the mobility of injected carriers. As follows from the results of measurements of the thermo-emf, the EPR, and the frequency dependence of the activation energy of conduction, the concentration of current carriers in the regions depends extremely little on the temperature. Since the conductivity of the polyconjugated macromolecules is known from the results of measurements in alternating current and is close to the pre-exponential factor σ_0, the microscopic mobility can be evaluated in regions of polyconjugated bonds. The estimates give a value of 10-100 $cm^3/V \cdot sec$, which confirm the applicability of the band model for the regions of polyconjugation. At the same time, the overall mobility is several orders of magnitude less; as obtained indirectly by substitution of the concentration of carriers in the expression for conductivity it is found to be between 10^{-2} and 10^{-6} $cm^2/V \cdot sec$ for different materials [3-7, 17, 34, 22]. Direct measurements of the overall mobility show that the mobility increases monotonically from 10^{-6} to 10^{-3} $cm^2/V \cdot sec$ with a growth in the regions of polyconjugation. As follows from what has been said, such low values of the effective mobility do not contradict the applicability of the band scheme for the regions of polyconjugation.

The second process — the transfer of a carrier from one region to another — has been studied far less. It is possible to evaluate only the energy side of this process from the difference of the activation energies of the overall conduction and that within the macromolecule. The tunnel mechanism for passing the barrier has been used in a number of papers [87-94]. However, in the majority of cases the idea of the tunnel passage of the barrier is inapplicable since a considerable activation energy of the transfer process is found experimentally.

At values of the activation energy of the order of 0.1-0.2 eV a transition analogous to the jump of a charge between point centers is probably possible. Consequently, attempts to predict the motion of an electron by means of the mechanism of the jumps that limit the mobility of the carriers is extremely interesting. According to Kventsel' [105], we may assume that the system considered consists of a number of polymer chains with conjugated bonds in which the electron moves along the chains, jumping from one link to another. Let the probability of such a jump within the macromolecule W_1 be far greater than the probability W_2 of a jump from one chain to another. Then, on the assumption that the jumps are absolutely random, we can describe the motion of the electron withing the framework of the problem of random walks [104]. In this case the motion of the electron can be described by a diffusion equation of the following form:

$$\frac{\partial F}{\partial t} = D_1 \frac{\partial^2 F}{\partial x^2},$$

where F(x, t) is the probability that the electron is present at point x at moment of time t, and $D_1 = W_1 l^2$ (where l is the length of a link). As the initial condition, $F(x, 0) = \delta(x + L)$ is selected.

Now making use of the inequality $W_1 \gg W_2$ we can write that $\left.\frac{\partial F}{\partial x}\right|_{-L} = D_1 \left.\frac{\partial F}{\partial x}\right|_{L} = 0$, where L and −L are the coordinates of the chain under consideration. These limiting conditions mean that the electron, having reached the end of a chain, most probably turns back and does not jump to a neighboring chain. The solution of the equation given, with these initial and boundary conditions, has the form

$$F(x, t) = \sum_{n=1}^{\infty} (-1)^n \cos \frac{\pi n x}{L} \exp\left(-\frac{\pi^2 D_1}{L^2} n^2 t\right).$$

The average time $\bar{t}$ required by the electron for its passage will be

$$t = \frac{\sum\limits_{n=1}^{\infty} \int\limits_0^{\infty} \exp\left(-\frac{\pi^2 D_1 n^2}{L^2} t\right) t\, dt}{\sum\limits_{n=1}^{\infty} \int\limits_0^{\infty} \exp\left(-\frac{\pi^2 D_1}{L^2} n^2 t\right) dt} = \frac{L^2}{\pi^2 D_1} \frac{\sum\limits_{n=1}^{\infty} \frac{1}{n^4}}{\sum\limits_{n=1}^{\infty} \frac{1}{n^2}} = \frac{L^2}{15 D_1} = \frac{N^2}{15 W_1},$$

where N is the number of links in the chain. The total time consumed in motion along the chain and in the jump of a carrier to another chain is $T = t + W_2^{-1}$. Then the effective diffusion coefficient

$$D_{\text{eff}} = W (2L)^2 = \frac{1}{15} \frac{W_1 W_2}{N^2 W_2 + 15 W_1} (2L)^2.$$

By making use of the Einstein relation we can obtain a formula for the mobility:

$$\mu = \frac{e\,(2L)^2}{kT} \frac{15 \overline{W}_1 \cdot \overline{W}_2}{N^2 W_2 + 15 W_1}.$$

The bar over the top denotes averaging over the various possible distances between the ends of neighboring chains. As can be seen from the formula, neglecting the time of passage along the chain leads to the mobility $\mu = e(2L)^2W_2/kT$. The dependence of the mobility on the temperature — the parameter that is most often studied experimentally — may be different since the dependence of W_1 and W_2 on the temperature is unknown.

REFERENCES

1. G. Bush and W. Winkler, Determination of the Characteristic Parameters of Semiconductors [Russian translation], IL, Moscow (1959).
2. A. Epstein and B. Wildi, J. Chem. Phys., 32:324 (1960); Khim. i Tekhnol. Polimerov, 1960(2):31.
3. E. I. Balabanov, E. L. Frankevich, and L. G. Cherkashina, Vysokomolek. Soedin., 5:1684 (1963).
4. A. A. Berlin, L. G. Cherkashina, E. L. Frankevich, E. I. Balabanov, and Yu. G. Aseev, Vysokomolek. Soedin., 6:832 (1964).
5. E. L. Frankevich, E. I. Balabanov, L. G. Cherkashina, Vysokomolek. Soedin., 6:1028 (1964).
6. A. M. Hermann and A. Rembaum, J. Appl. Phys., 37:3642 (1966).
7. H. Pohl and E. Engelhardt, J. Phys. Chem., 66:2121 (1962).
8. H. Pohl, Proceedings of the 5th Conference on Carbon, Pergamon Press, Oxford (1963), p. 113.
9. G. H. Heilmeier, G. Warfield, and S. E. Harrison, Phys. Rev. Letter, 8:309 (1962).
10. G. Delacote and H. Shott, Phys. Stat. Sol., 2:1460 (1962).
11. G. H. Heilmeier and S. E. Harrison, Phys. Rev., 132:2010 (1963).
12. G. H. Heilmeier, G. Warfield, and S. E. Harrison, J. Appl. Phys., 34:2278 (1963).
13. F. Winslow, W. Baker, and W. Jager, J. Am. Chem. Soc., 77:4751 (1955).
14. P. Fielding and F. Gutman, J. Chem. Phys., 26:411 (1957).
15. G. Garret, Rad. Res. Suppl., 2:340 (1960).
16. R. M. Voitenko and É. M. Raskina, Dokl. Akad. Nauk SSSR, 136:1137 (1961).
17. A. V. Airapetyants, R. M. Voitenko, B. É. Davydov, and B. A. Krentsel', Dokl. Akad. Nauk SSSR, 148:605 (1963).
18. R. M. Vlasova and A. V. Airapetyants, Élektrokhimiya, 1:962 (1965).
19. R. M. Vlasova and A. V. Airapetyants, Fiz. Tverd. Tela, 7:3079 (1965).
20. N. A. Bakh, V. D. Bityukov, A. V. Vannikov, and A. D. Grishina, Dokl. Akad. Nauk SSSR, 144:135 (1962).
21. A. V. Vannikov, V. D. Bityukov, and N. A. Bakh, in: Chemical Properties and Modification of Polymers [in Russian], "Nauka," Moscow (1964), p. 41.
22. L. I. Boguslavskii and L. S. Stil'bans, Vysokomolek. Soedin., 6:1802 (1964).
23. E. L. Frankevich, L. I. Busheva, E. I. Balabanov, and L. G. Cherkashina, Vysokomolek. Soedin., 6:1028 (1964).
24. H. A. Pohl and G. A. Bornman, Khim. i Tekhnol. Polimerov, 1961:1181.
25. H. A. Pohl, G. A. Bornman, and W. Itoh, in: Organic Semiconductors, J. J. Brophy and J. W. Buttney (eds.), Macmillan, New York (1962), p. 142.

26. W.D. Brennan, J.J. Brophy, and G. Schonhorn, ibid., p. 159.
27. D.D. Eley and G.D. Parfitt, Trans. Faraday Soc., 51:1529 (1955).
28. N.B. Hannay (ed.), Semiconductors, Reinhold, New York (1959).
29. A.A. Berlin, L.I. Boguslavskii, R.Kh. Burshtein, N.G. Matveeva, A.I. Sherle, and N.A. Shurmovskaya, Dokl. Akad. Nauk SSSR, 136:1127 (1961).
30. C.M. Higgins and A.H. Sharnaugh, J. Chem. Phys., 38:393 (1963).
31. A.V. Vannikov, L.P. Sidorova, V.I. Yakovenko, and N.A. Bakh, Élektrokhimiya, 2:1474 (1966).
32. L.I. Boguslavskii, A.I. Sherle, and A.A. Berlin, Zh. Fiz. Khim., 38:1118 (1962).
33. I. Storbeck and M. Starke, Ber. des Bunsenges. Phys. Chem., 89:343 (1965).
34. N.A. Bakh, A.V. Vannikov, A.D. Grishina, and S.V. Nizhnii, Usp. Khim., 34:1733 (1965).
35. A.V. Vannikov and N.A. Bakh, Élektrokhimiya, 1:617 (1965).
36. S. Tanaka and H.Y. Fan, Phys. Rev., 132:1516 (1963).
37. M. Pollak, Phys. Rev., 133(2A):564 (1964).
38. L.I. Boguslavskii and L.S. Stil'bans, Dokl. Akad. Nauk SSSR, 147:1114 (1963).
39. L.I. Boguslavskii, Zh. Fiz. Khim., 39:748 (1965).
40. L.I. Boguslavsky and A.V. Vannikov, J. Electrochem. Soc., 111:755 (1964).
41. G.I. Skanavi, The Physics of Dielectrics [in Russian], Gostekhteoretizdat, Moscow (1949).
42. C. Koops, Phys. Rev., 83:121 (1951).
43. A.A. Garev and A.I. Piryazeva, Zh. Teoret. Fiz., 21:1469 (1951).
44. É.M. Trukhan, Fiz. Tverd. Tela, 4:3496 (1963).
45. R. Petrits, Phys. Rev., 104:1508 (1956).
46. C.A. Mead, Phys. Rev., 128:2088 (1962).
47. H. Hirose and Y. Wada, Jap. J. Appl. Phys., 4:639 (1965).
48. J.J. Brophy, J. Appl. Phys., 33:114 (1962).
49. R.G. Kepler, Phys. Rev., 119:1226 (1960).
50. O.H. Le Blanc, J. Chem. Phys., 33:626 (1960).
51. W.E. Spear, Proc. Phys. Soc., B70:669 (1957).
52. W. Helfrich and H. Mark, Z. Physik, 166:370 (1962).
53. P.N. Keating and A.C. Papadakis, Proceedings 7th International Conference on the Physics of Semiconductors, Paris (1964), p. 519.
54. M. Silver, D. Olness, M. Swicord, and R. Jarning, Phys. Rev. Letters, 10:12 (1963).
55. D. Hoesterey and G. Letston, J. Phys. Chem. Solids, 24:1609 (1963).
56. D. Hoesterey and G. Letston, J. Chem. Phys., 41:675 (1964).
57. W. Mehl and N. Wolf, J. Phys. Chem. Solids, 25:1221 (1964).
58. R. Raman and S.P. McGlynn, J. Chem. Phys., 40:515 (1964).
59. R. Raman, L. Azarraga, and S.P. McGlynn, J. Chem. Phys., 41:2516 (1964).
60. E.L. Frankevich and E.I. Balabanov, Fiz. Tverd. Tela, 7:710 (1965).
61. E.L. Frankevich and E.I. Balabanov, Phys. Stat. Sol., 14:523 (1966).
62. A.V. Vannikov, V.I. Zolotarevskii, and D.I. Naryadchikov, Élektrokhimiya, 3:1379 (1967).
63. A.V. Vannikov, L.I. Boguslavskii, and V.B. Margulis, Fizika i Tekhnika, Poluprovodnikov, 1:935 (1967).
64. K.K. Thornber and C.A. Mead, J. Phys. Chem. Solids, 26:1488 (1965).

65. R. G. Kepler, in: Organic Semiconductors, J. J. Brophy and J. W. Buttney (eds.), Macmillan, New York (1962), p. 1.
66. P. T. M. Silver, M. Swicord, and R. C. Jarning, J. Phys. Chem. Solids, 23:419 (1962).
67. J. Gränacher, Sol. State Comm., 2:365 (1964).
68. O. H. Le Blanc, in: Organic Semiconductors, J. J. Brophy and J. W. Buttney (eds.), Macmillan, New York (1962), p. 21.
69. O. H. Le Blanc, J. Chem. Phys., 35:1275 (1961).
70. M. Silver, in: Organic Semiconductors, J. J. Brophy and J. W. Buttney (eds.), Macmillan, New York (1962), p. 21.
71. A. R. Adams and W. E. Spear, J. Phys. Chem. Solids, 25:1113 (1964).
72. R. Silbey, J. Jorner, S. A. Rice, and M. T. Vala, J. Chem. Phys., 42:733 (1965).
73. W. Bepler, Z. Physik, 185:507 (1965).
74. M. C. Tobin and D. P. Sitzer, J. Chem. Phys., 42:3652 (1965).
75. O. H. Le Blanc, J. Chem. Phys., 37:916 (1962).
76. J. Yamashita and T. Kurosawa, J. Phys. Soc. Japan, 15:802 (1960).
77. K. Masuda and J. Yamaguchi, J. Phys. Soc. Japan, 19:1130 (1964).
78. A. V. Vannikov, Fiz. Tverd. Tela, 9:1367 (1965).
79. H. Fröhlich and G. L. Sewell, Proc. Phys. Soc., 74:643 (1959).
80. M. Goppert-Mayer and A. J. Sklar, J. Chem. Phys., 6:645 (1938).
81. S. H. Glarum, J. Phys. Chem. Solids, 24:1577 (1963).
82. E. O. Förster, Electrochem. Acta, 9:1319 (1964).
83. R. M. Glaeser and R. S. Berry, J. Chem. Phys., 44:3797 (1966).
84. G. Delacote and M. Schott, Solid State Comm., 4:177 (1966).
85. S. I. Kubarev and I. D. Mikhailov, Zh. Éksperim. Teoret. Khim., 1:229 (1965).
86. S. I. Kubarev and I. D. Mikhailov, Zh. Éksperim. Teoret. Khim., 1:488 (1965).
87. V. A. Kargin, A. V. Topchiev, B. A. Krentsel', L. S. Polak, and B. É. Davydov, Zh. Vses. Khim. Obshchestva im. D. I. Mendeleeva, 5:507 (1960).
88. I. M. Kustanovich, I. I. Patalakh, and L. S. Polak, Kinetika i Kataliz, 6:167 (1963).
89. I. M. Kustanovich, I. I. Patalakh, and L. S. Polak, Vysokomolek. Soedin., 6:197 (1964).
90. S. A. Nizova, I. I. Patalakh, and L. S. Polak, Dokl. Akad. Nauk SSSR, 153:144 (1963).
91. D. Eley and D. Spivey, Trans. Faraday Soc., 56:1432 (1960).
92. D. Eley and D. Spivey, Trans. Faraday Soc., 57:2286 (1961).
93. F. Cardwell and D. Eley, Disc. Faraday Soc., 28:54 (1959).
94. D. Eley, Research, 12:293 (1959).
95. H. Pohl, G. Cappos, and G. Jages, J. Polymer Sci., A140:547 (1963).
96. N. Juster, J. Chem. Educ., 40:547 (1943).
97. Y. Okamoto and W. Brenner, Organic Semiconductors, Reinhold, New York (1964).
98. H. Pohl, Proceedings of the 4th Conference on Carbon, New York (1960), p. 241.
99. P. Waters, Proceedings of the 5th Conference on Carbon, New York (1964), p. 131.
100. L. S. Stil'bans and L. D. Rozenshtein, in: Electrical Conductivity of Organic Semiconductors [Russian translation], IL, Moscow (1963), p. 5.
101. V. G. Levich, V. S. Markin, and Yu. G. Chirkov, Dokl. Akad. Nauk SSSR, 149:894 (1963).

102. J. Koutecky and R. Zahradnik, Abstracts of a Conference on Semiconductor Physics, Prague (1960), p. 122.
103. H. Kuhn, J. Chem. Phys., 16:840 (1948).
104. C. S. Chandrasekhar, Stochastic Problems in Physics and Astronomy, Rev. Mod. Phys., 15:2 (1943).
105. G. F. Kventsel', Zh. Éksperim. Teoret. Khim., 1:826 (1965).

CHAPTER V

THE SURFACE OF ORGANIC SEMICONDUCTORS

There is no doubt that the surface properties of organic semiconductors are extremely important both for understanding the processes taking place at the boundary of separation of two phases and for more complete understanding of the processes taking place in the bulk of the semiconductor. The influence of the state of the surface and the processes taking place at the surface on the bulk properties is particularly strong when the semiconductors have a low intrinsic conductivity [1]. In the adsorption of gases on the surface of a semiconductor it has been found experimentally that some molecules are capable of increasing the concentration of electrons in the surface layer and others of decreasing it. The process of adsorption may be considered in two ways: either as a creation of donor (acceptor) impurities at the surface [2, 3] or as the formation of charge-transfer complexes [4]. The appearance of additional carriers on the adsorption at the surface of an acceptor can be represented by the following equations for n- and p-type semiconductors:

$$\left.\begin{array}{l} \frac{1}{2}A_2 + e \rightarrow A_{\text{ads}} \\ \frac{1}{2}A_2 \rightarrow A_{\text{ads}} + p. \end{array}\right\} \qquad (1)$$

According to the scheme given, the process of chemisorption can be traced by the change in the surface conductivity. The adsorption of an acceptor gas on a p-type semiconductor increases the conductivity and the adsorption of a donor gas decreases it. The extremely high sensitivity of the conductivity to gases adsorbed from the surface can be used in practice for analytical purposes [5].

INFLUENCE OF THE ADSORPTION OF GASES ON THE SURFACE CONDUCTIVITY AND THE THERMO-EMF

The conductivity of organic semiconductors in which the concentration of inherent current carriers is extremely low depends greatly on the state of the surface and on the nature of the contacts by means of which the specimens are connected to the measuring circuit. Depending on the ratio of the work functions of the contact and the organic semiconductor and also the distortion of the bands of the semiconductor at the point of contact, the injection of holes or electrons into the semiconductor may be observed.

At low voltages imposed on the sample, Ohm's law is always observed, and this is valid until the concentration of injected carriers becomes comparable with the concentration of intrinsic carriers in the semiconductors.

The theory of space-charge-limited currents gives an exhaustive description of the processes at the metal–semiconductor surface. In the case of electrolytic contacts some estimates have been made by Kallmann and Pope [11]. In particular, for an anthracene electrode placed in a solution the injection of holes was observed only where the following inequality was valid

$$I < \varphi + E, \tag{2}$$

where φ is the work function of an electron from the contact material, I is the ionization potential of anthracene in the solid state, and E is the additional energy that can be obtained from the electric field or the energy of the light quanta.

When gases are adsorbed on the surface of a semiconductor, the adsorbed molecules form dipoles. The partial transfer of charge from an adsorbed molecule to the semiconductor may (it is true, to a very rough approximation) be regulated by relation (2). So far as concerns the results of experiments, electrical conductivity is very sensitive to the state of the surface. Consequently, workers frequently come to extremely contradictory conclusions. In a study of the conductivity of anthracene in a cell of the layer type using a guard ring, no influence of oxygen on the photocurrent was observed [6]. The same result was obtained in an investigation of the influence of oxygen on the conductivity of metal-free phthalocyanine [7]. Examples of the opposite situation can also be given. An influence of oxygen on the conductivity of copper phthalocyanine with the use of a guard ring has been observed [8]. The influence of various gases on the conductivity of anthracene and p-chloranil—amine complexes has been studied in a cell of the layer type [9, 10]. In a study of the influence of iodine on the conductivity of a single crystal of anthracene it was found that the conductivity in an atmosphere of iodine rises with an increase in its vapor pressure. It is possible that an anthracene—iodine complex is formed on the surface of the crystal, the surface being a donor and the adsorbed iodine an acceptor of electrons.

In Fig. 30 the vapor pressure of the iodine has been plotted as the ordinate and the ratio of the specific resistances of the specimen before and after treatment in iodine vapor ρ_{init}/ρ_{final} as the abscissa. Even at an iodine vapor pressure of 5×10^{-1} mm Hg the bulk conductivity rises several hundred times.

The sensitivity of the electrical conductivity to iodine vapor depends on the direction of the crystallographic axis. Plates cut perpendicular to the AB plane are sensitive at pressures as low as 10^{-4} mm Hg. Plates parallel to the AB plane are insensitive in the low-pressure region but at higher pressures of iodine the conductivity increases rapidly. It has been impossible to obtain a change in the conductivity of stilbene and naphthalene under the action of iodine vapor. This can be explained by the fact that the formation of a charge-transfer complex is possible between iodine and ferrocene and between iodine and pyrene but is energetically unfavorable in the case of naphthalene and stilbene [see inequality (2)].

The influence of water vapor on the surface conductivity of a single crystal of anthracene has also been studied [12]. On a

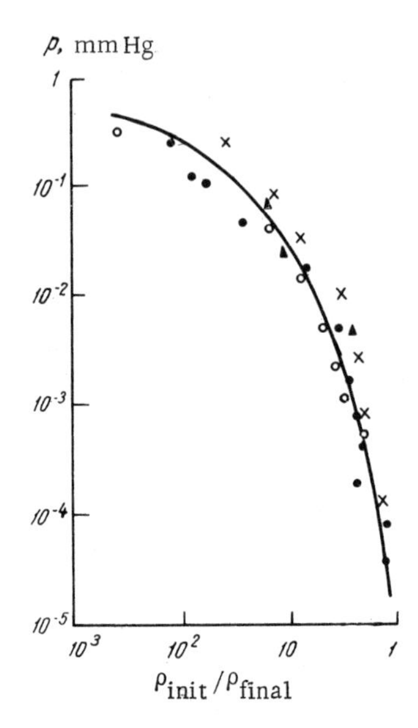

Fig. 30. Influence of the iodine vapor pressure p on the resistance of anthracene [in relative units (ρ_{init}/ρ_{final})].

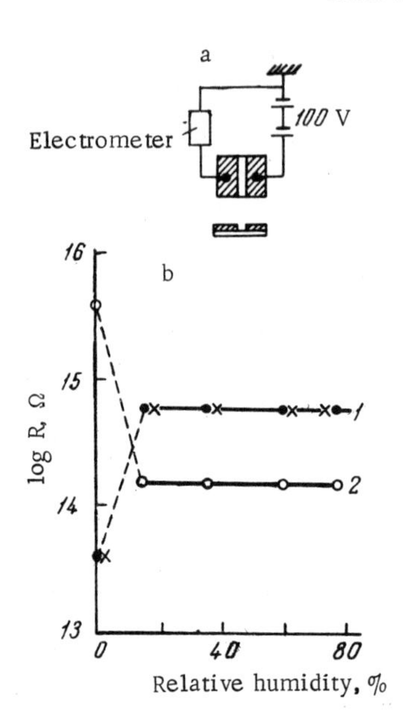

Fig. 31. Change in the surface conductivity of anthracene in moist air for a fresh surface (1) and one corroded in air (2).

clean surface, with an increase in the relative humidity to 20% the resistance falls by almost two orders of magnitude. Curiously, a surface subjected to treatment in water vapor for months exhibits the opposite effect (Fig. 31). This is explained by the formation on the surface of the anthracene of derivatives of it containing oxygen, such as, for example, anthranol.

TABLE 11. Change in the Specific Resistance ρ of p-Chloranil in Various Media [9]

Medium	ρ_{init}, Ω·cm	mm Hg	ρ_{init}/ρ_{final}
Ammonia	$3 \cdot 10^{16}$	20	1.7
Trimethylamine	$1.5 \cdot 10^{15}$	20	12
Methylamine	$4.0 \cdot 10^{15}$	20	24
Triethylamine	$1.7 \cdot 10^{16}$	20	96
Diethyl phosphoro-chloridate	$3 \cdot 10^{15}$	0.46	0.32
Iodine	$3 \cdot 10^{15}$	0.25	0.20

TABLE 12. Change in the Activation Energy E of the Electrical Conduction of p-Chloranil in Various Media [9]

Medium	p, mm Hg	E, eV	Medium	p, mm Hg	E, eV
Methylamine	0	0.63	Ammonia	0	0.85
	2	0.51		27	0.79
	13	0.41		70	0.64
	41	0.29		160	0.56
				505	0.50
Trimethylamine	0	0.85			
	2	0.55			
	14	0.44			
	43	0.44			

The influence of various gases on electrical conductivity has been observed in the treatment of p-chloranil in the vapors of substances possessing both donor and acceptor properties. Table 11 gives the results of the treatment of p-chloranil in various media. In this case iodine does not increase but decreases the conductivity of the crystal. As a rule the sign of the effect depends on whether the surface of the semiconductor acts as a donor or acceptor of electrons in relation to the adsorbed gas. If the complex formed on the surface repels electrons, then in the case of electronic conductivity this leads to an increase in specific resistance. The opposite effect will be observed in the case of a hole conductor. Consequently, knowing the nature (donor or acceptor) of the adsorbed gas, it is usually possible to establish whether we are dealing with n- or p-type of conduction.

With the same p-chloranil as an example, it is possible to convince oneself that not only the resistance but also the thermal activation energy of conduction $\Delta E(\rho = \rho_0 \exp E/kT)$ depends on the state of the surface of the material investigated (Table 12).

Polymeric semiconducting materials, particularly powders and films, which possess a developed surface, are also subject to the influence of adsorbed gases. It is known that polyacrylonitrile, which is a hole semiconductor in the air, changes not only the magnitude but also the type of conduction when its surface is degassed [7]. Since polyphenylacetylene also changes the magnitude of its electrical conductivity on the adsorption of oxygen, by making use of equation (1) Balabanov et al. [14], knowing the amount of ad-

TABLE 13. Change in the Type of Conduction and the Thermo-emf of Polyphthalocyanine in Various Media [13]

Medium	Type of conduction	Thermo-emf, μV/deg
Water	n	59
Hydrogen sulfide	n	30
Bromine	p	1
Ozone	p	5
Vacuum	p	12–18

sorbed gas, were able to evaluate the concentration of carriers in the polymer, which proved to be 3×10^{18} cm^{-3}.

Water vapor, hydrogen sulfide, bromine, and ozone change the type of conduction and the magnitude of the thermo-emf of low-temperature polyphthalocyanine (Table 13).

It must be observed that the adsorption and desorption of gases in various materials compressed in the form of tablets takes tens of hours. Water vapor and oxygen are frequently adsorbed so strongly that they are removed only by heating for many hours. A mass-spectrometric investigation has shown that when the samples are subjected to vacuum treatment not only the water vapor and the oxygen are liberated but also products of the degradation of the terminal groups of the polymers [14].

However, scheme (1) is not always satisfied experimentally. In particular, in the adsorption of oxygen on polyphenylacetylene [15] the oxygen suppresses the electrical conductivity although the polymer has hole conduction. This is explained by the fact that the oxygen obviously creates new centers of recombination which lead to a decrease in the lifetime of the majority carriers. There is no doubt that in order to be able to judge correctly the influence of adsorbed gases further information is necessary on the state of the surfaces.

Unfortunately, extremely few investigations have been devoted to the state of the surface of organic semiconductors. The effect of a field on the surface photopotential in atmospheres of various donor and acceptor gases has been measured on a film of copper α-phthalocyanine [16]. However, although an influence of the gases on the conductivity was in fact observed, it was not possible to obtain the usual curve with a minimum surface conductivity because of the high density of the acceptor states, the concentration of which (by a rough estimate) was not less than 10^{13} cm^{-2}. These surface states apparently owe their origin to oxygen. Similar results have been obtained for anthracene, where the concentration of traps was found to be 10^{17} cm^{-3} [17, 18].

CHANGE IN THE WORK FUNCTION ON THE ADSORPTION OF GASES ON ORGANIC SEMICONDUCTORS

The magnitude of the change in the work function of an electron from an organic semiconductor on adsorption can be used as a criterion of the donor or acceptor properties of a gas. A comparison of the change in the Volta potential and the relative change in the conductivity shows how well relation (1) is satisfied for various gases. Such a check has been carried out for a number of gases on radiation-thermally modified polyethylene (RTMP) which had a conductivity of $10^{-13}\,\Omega^{-1}\cdot\text{cm}^{-1}$ and an activation energy of conduction of 1.4 eV. According to scheme (1), an acceptor gas should increase the conductivity of the film since on adsorption the concentration of holes should increase. Table 14 [18] shows the

TABLE 14. Change in the Work Function $\Delta\varphi$ and the Relative Electrical Conductivity of RTMP in an Atmosphere of Various Gases

Medium	$\Delta\varphi$	σ/σ_0
Oxygen	0.23	0.7
	0.15	1.2
Hydrogen	0.1	7.5
Iodine	1.0	7
Water	0.25	4
N,N-Dimethyl-n-octylamine	–0.1	0.3

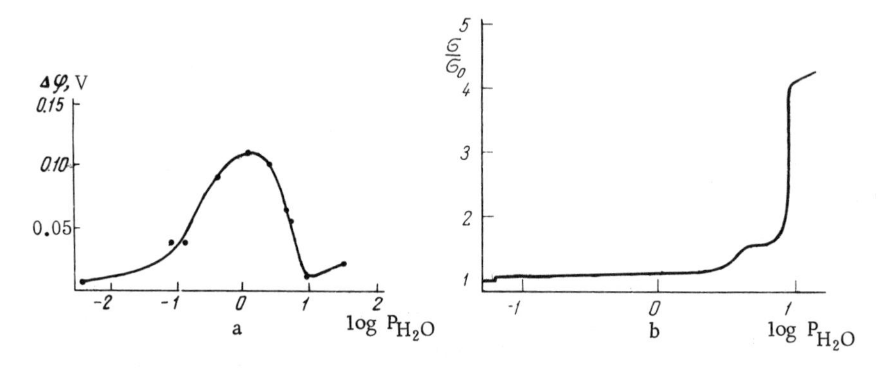

Fig. 32. Change in the Volta potential (a) and surface conductivity (b) of radiation-modified polyethylene on the adsorption of water.

change in the work function and the surface conductivity on the adsorption of oxygen, hydrogen, iodine, water, and N,N-dimethyl-n-octylamine.

As can be seen from Table 14, only the amine decreases the work function and is, therefore, a donor gas. All the other substances are acceptors to a greater or smaller extent with respect to the semiconductor film and must increase its electrical conductivity. However, in a consideration of the action of oxygen it can be seen that, depending on the state of the film, it may either increase or decrease its electrical conductivity: oxygen admitted to a carefully pumped-out surface decreases its electrical conductivity. When the same amount of gas (10 mm Hg) is admitted to an already-oxidized surface, an increase in the conductivity of the film is observed.

A more many-sided investigation of the action of gases on the surface conductivity of a polymer frequently permits an explanation of extremely fundamental features of the action of adsorbed gases [19]. Figure 32 shows the change in the Volta potential and the surface conductivity of RTMP on the adsorption of water vapor. It can be seen from the upper part of the figure that the Volta potential passes through a minimum, which can be explained by the formation of polymolecular layers at $\theta = 1$. It is interesting that the surface conductivity rose sharply at the same pressure. Thus, at high pressures of water vapor the conductivity of a polymolecular film of water is observed.

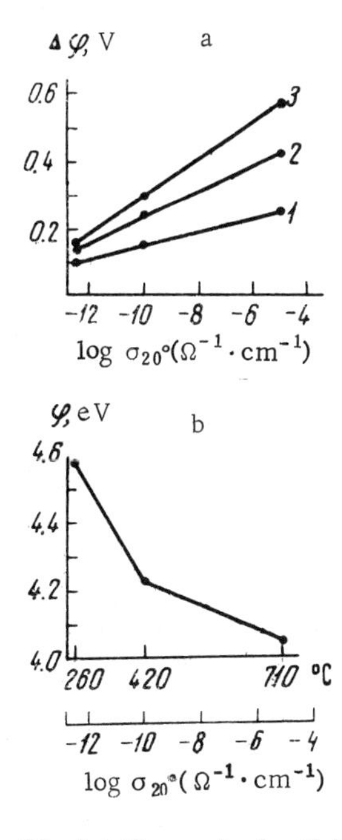

Fig. 33. (a) Change in the Volta potential on the adsorption of 10 mm Hg of oxygen on RTMP of different conductivities, and (b) the dependence of the work function on the temperature of the thermal treatment of the sample. 1) 25°C; 2) 100°C; 3) 200°C.

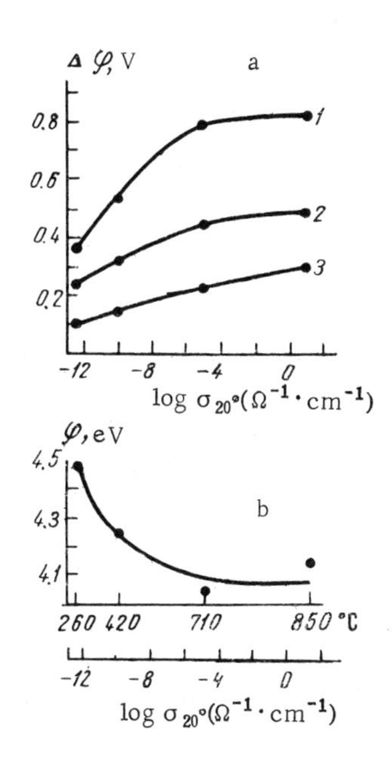

Fig. 34. (a) Change in the Volta potential on the adsorption of iodine on RTMP of different conductivities, and (b) dependence of the work function on the temperature of the thermal treatment of the sample. 1) 25°C; 2) 100°C; 3) 200°C.

The same effect, but less sharply, was found with the adsorption of iodine. Particularly interesting are the results of an investigation of the dependence of the magnitude of the dipole formed by adsorption on the conductivity and the work function of an electron from the surface of a degassed organic semiconductor [20]. This relationship was observed for the adsorption of such acceptors as oxygen and iodine on a film of radiation-modified polyethylene the conductivity of which varied between 10^{-13} and $10\ \Omega^{-1}\cdot cm^{-1}$ for the four samples investigated. Oxygen was adsorbed at a pressure of 10 mm Hg and iodine at 10^{-1} mm Hg. This ensured the maximum degree of coverage of the surface in all the experiments.

The adsorption of oxygen is basically chemical. This can be seen clearly from the temperature dependence of the effect, which increases with a rise in the temperature (Fig. 33). Conversely, the change in the work function on the adsorption of iodine carried out at various temperatures shows the predominance of physical adsorption (Fig. 34). A more detailed description of the features of these two processes is given by Boguslavskii and Margulis [18, 19]. As can be seen from the figures, the magnitude of the change in the Volta potential rises with an increase in the conductivity and a decrease in the work function of an electron from the degassed polymer. This result proved equally correct for chemisorption and for physical adsorption.

CHANGE IN THE PHOTOELECTRIC SENSITIVITY ON ADSORPTION

The study of the sensitization effect is important for explaining the role of the surface states of a semiconductor. As is known, this effect consists in the sensitization of the photoelectric response of the semiconductor by dyes in the region of the spectrum corresponding to the optical absorption of the molecule of the dye. It has been established [21] that the presence in the semiconductor of photoelectric conductivity in the long-wave region of the spectrum is of great importance for sensitization.

Various actions having a long-wave photoelectric sensitivity are strongly reflected in the sensitization effect. The adsorption of electronegative gases and vapors such as oxygen, iodine, quinone, and bromine on various semiconductors both changes the region of the inherent photoelectric sensitivity and simultaneously increases the dye-sensitized photoeffect by tens of times. Since the molecules of the gases and vapors mentioned possess high electron affinities, it may be assumed that trapping levels for electrons on the surface of the semiconductor, arising on the adsorption of electronegative molecules, are necessary for sensitization. Thus, the state of the surface has a decisive role for the appearance of the sensitization effect and can be studied by means of it.

At the present time, two mechanisms of sensitization have been proposed (Fig. 35). According to the first mechanism, the

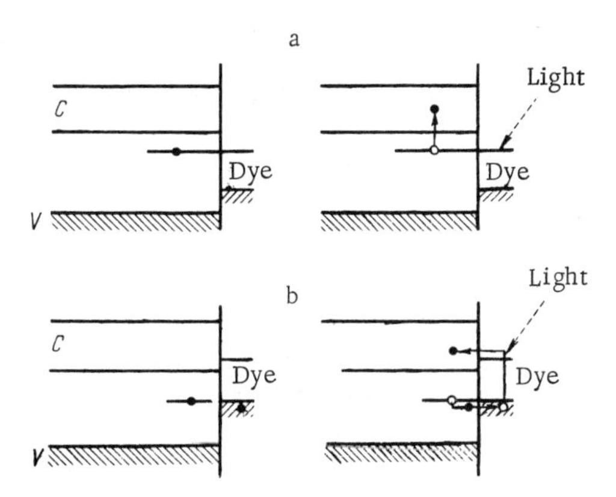

Fig. 35. Resonance mechanism (a) and charge-transfer mechanism (b) of the sensitization of a semiconductor by a dye.

absorption of a photon in the molecule of the adsorbed dye causes the expulsion of an electron into an excited level which lies above the bottom of the conduction band of the semiconductor. Then the electron falls into the conduction band of the semiconductor, which leads to the appearance of a photocurrent. An electron present in the trapping level in the semiconductor passes to the molecule of the dye with its subsequent regeneration [22, 23]. Another point of view is that the energy of the light quantum absorbed by the adsorbed molecule of the sensitizer dye is transferred to the semiconductor by a resonance mechanism. Under the action of this energy an electron from the local surface level passes into the conduction band, which leads to the appearance of photoconductivity. In this case it is not necessary that the excited level of the dye molecule should be above the bottom of the conduction band [21]. At the present time it is not possible to make a definitive choice between the two mechanisms of sensitization that have been described.

The sensitization of the photoeffect by dyes has recently been observed also in anthracene [24], poly(copper phenylacetylide) [25], and the leuco form of malachite green [26].

INFLUENCE OF THE ADSORPTION OF GASES ON THE EPR SIGNAL

The formation of charge-transfer complexes on the adsorption of gases on the surface can be established by the change in the EPR signal. Usually, to increase the observed effect, substances are studied in the powdered state. This, however, presents a certain disadvantage in view of the fact that the measurement of the surface of the samples does not take place simultaneously and the data obtained from the EPR signal are not comparable with the adsorption measurements.

To some degree, the disadvantage mentioned has been taken into account by studying the EPR spectra of a single crystal of p-chloranil in an atmosphere of ammonia [9]. No EPR signal was detected on a single crystal weighing 0.12 g the surface of which was ~ 1 cm^2; however, it did appear when the crystal was comminuted. The signal obtained showed the presence of 10^{14}–10^{15} unpaired spins in a sample with $g = 2.0043$ and a line width of 11 Oe. The absence of a signal in the case of the single crystal, on the assumption that the adsorption was in the form of a monolayer, leads to the conclusion that less than 0.14 of all the adsorbed molecules had an unpaired spin.

The EPR method has been used to investigate the formation of paramagnetic complexes in the reaction of oxygen with disperse samples of magnesium phthalocyanine (β-modification) at high temperatures [27]. It was found that the concentration of paramagnetic complexes increases exponentially with a rise in the temperature ($E = 10$ kcal/mole). On the basis of kinetic measurements of the accumulation of the complexes and their decomposition when the oxygen was pumped off, it was suggested that the growth of the EPR signal of the paramagnetic complexes is due to the thermal energy of activation of the centers capable of complex-formation. The diffusion coefficient of the oxygen in a molecular crystal of magnesium phthalocyanine was evaluated from the kinetic curves of the formation of the paramagnetic centers. It proved to be 3×10^{-3} cm^2/sec, and the activation energy of this process was 4.0 ± 1.6 kcal/mole.

An investigation of metal-free phthalocyanine [8] in the form of a powder in an atmosphere of oxygen led to an increase in the

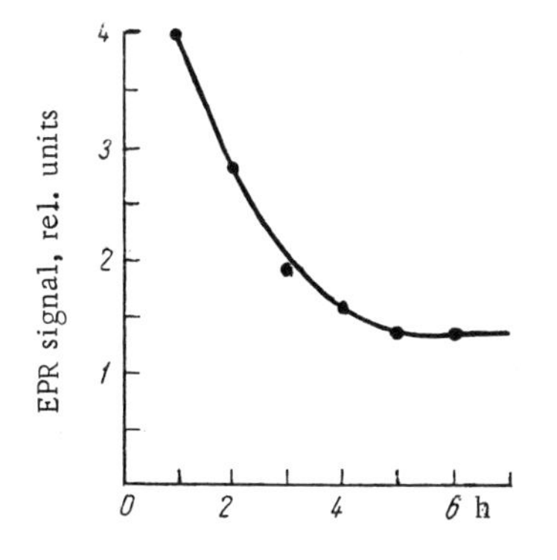

Fig. 36. Change in the EPR signal with time during the heating of metal-free phthalocyanine in hydrogen.

EPR signal. The line width was $\Delta H = 54 \pm 0.5$ Oe and it changed somewhat on heating, while g was 2.0024 ± 0.003 and remained constant.

When metal-free phthalocyanine was heated in an atmosphere of hydrogen, the EPR signal decreased (Fig. 36). The observed effect was irreversible. The fact that metal-free phthalocyanine is a catalyst for the formation of water from its elements may help to explain this result [28]. Consequently, heating the sample in an atmosphere of hydrogen led to the freeing of the surface from adsorbed oxygen and to a decrease in the EPR signal. The observed unpaired spins can be explained by the presence of reversibly adsorbed oxygen which, in this case, plays the role of a free radical.

The formation of free radicals on the reversible adsorption of oxygen has also been noted in the case of some biological materials [29]. Although the shape of the signal was different in the two cases (in the case of phthalocyanine the signal is symmetrical and approximately twice as narrow) it is not excluded that the centers of adsorption may have the same electronic nature in both cases.

A completely different effect is given by the adsorption of oxygen on polyphthalocyanine [30]. In an atmosphere of oxygen the EPR signal disappears.

Before being placed in an atmosphere of oxygen, an EPR signal 2.4 Oe broad corresponding to a concentration of spins of 2.5×10^{17} g^{-1} was observed on a sample of polyphthalocyanine (both in vacuum and in an atmosphere of hydrogen).

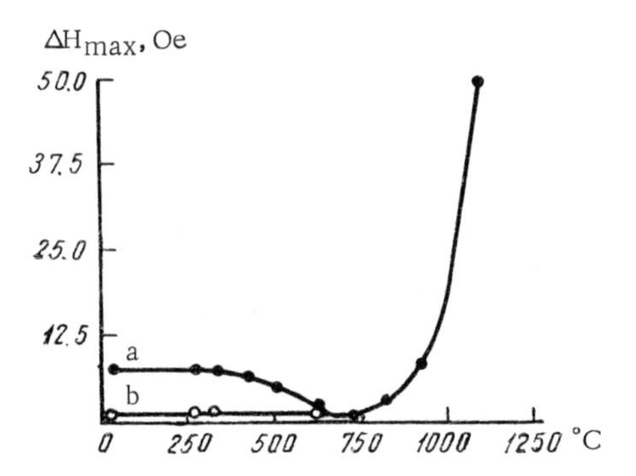

Fig. 37. Reversible adsorption of oxygen on modified polyethylene observed from the change in the line widths as a function of the temperature of thermal treatment of the samples. a) Pumping out for 2 h, b) 30 h.

The reversible nature of the adsorption of oxygen was established from the change in the EPR signal in samples of radiation-thermally modified polyethylene [32]. It was found that the concentration of paramagnetic centers for samples treated at a temperature above 600°C decreased greatly in the presence of oxygen. This effect depends on the degree of thermal treatment and is strictly reversible. Figure 37 shows the line widths as a function of the conditions under which the oxygen was removed for various samples.

The dependence of ΔH of the samples after pumping out for 2 h at 10^{-3} mm Hg on the temperature of the thermal treatment of the samples gives a curve with a sharp minimum. This relationship is analogous to those that are obtained on charcoals and other pyrolysis products [33-35]. However, with more careful pumping out of the oxygen continuously for 24 h at a pressure of 10^{-5}–10^{-6} mm Hg a line width of 0.5 Oe was obtained for all samples, beginning with the irradiated but nonpyrolyzed sample of polyethylene.

Thus, broadening of the line to 7.5 Oe takes place through the interaction of the paramagnetic centers with oxygen molecules. The adsorption of one molecule of oxygen leads to the reversible destruction of approximately two paramagnetic centers [36, 37].

The adsorption of hydrogen on copper polyphthalocyanine and its influence on the EPR signal has also been considered in an investigation of the ortho–para transition of hydrogen [30]. A hydrogen atmosphere does not affect the intensity of the narrow signal but leads to a considerable decrease in the width of the signal which can be ascribed to the paramagnetism of the chelate copper atom. Consequently it may be concluded that on adsorption onto polyphthalocyanine the hydrogen atom adds to the copper atom.

Thus, generally speaking, the adsorption of gases on the surface of organic semiconductors affects all their parameters. The magnitude and sign of the conduction, the thermo-emf, the Hall effect, and the EPR signal — all these depend on the nature and pressure of the adsorbed gas.

BOUNDARY OF SEPARATION BETWEEN AN ORGANIC SEMICONDUCTOR AND AN ELECTROLYTE

As has been shown, the resistance, thermo-emf, work function, and the other parameters of high-ohmic organic semiconductors depend strongly on the state of their boundary of separation with the gas phase. Recently, the investigation of the boundary of separation between organic semiconductors and electrolytes has led to a number of interesting results, which were obtained in a study of the processes of charge transfer and electrode reactions on "organic insulators" [38]. Under organic insulators we must also include organic semiconductors which have a high specific resistance and a broad forbidden gap.

The majority of the results given below were obtained in the study of single crystals of anthracene immersed in electrolytes of various compositions. When an external voltage is applied to an anthracene electrode, initially Ohm's law is observed for the volt–ampere curve. With sufficient accuracy it can be stated that the strength of the field at any point of the volume is connected with the applied voltage by the relation

$$E(x) = \frac{v}{d}, \tag{3}$$

and the current flowing through the sample can be defined as

$$i = en\mu \frac{v}{d}. \tag{4}$$

Equation (4) is valid as long as thermal equilibrium at the contact is observed. When the density of the carriers injected from the contact becomes comparable with the density of the carriers generated thermally, deviations from Ohm's law begin to be observed because of the formation of a space charge in the sample. Then a square or more pronounced dependence of the current on the applied voltage may be observed in connection with the presence of traps for the carriers in the crystal. At some particular voltage, the rise in current ceases [39]. The latter phenomenon has been studied in detail and has been given an explanation in a number of papers by Mehl [40–42]. The boundary of separation between anthracene and an electrolyte in these investigations was considered in the same way as is done in the electrochemistry of semiconductors [43].

The Fermi levels of the solution and the semiconductor coincide and the bending of the band at the boundary of separation between the anthracene and the electrolyte is determined by the redox potential of this system in accordance with the relations

$$\begin{aligned} \Delta E_s^n &= E_n + e(E_{ro} - E_0) \\ \Delta E_s^p &= E_p - e(E_0 - E_{ro}). \end{aligned} \tag{5}$$

In these equations ΔE_s^n is the bending of the conduction band, ΔE_s^p is the bending of the valence band, E_0 is the potential of the plane bands, E_{ro} is the redox potential, and E_p and E_n are the distances from the Fermi level for the valence band and the conduction band respectively.

Since the rate of occurrence of a redox reaction at the surface of the insulator is limited by the concentration of current carriers at the anthracene–electrolyte boundary [44–47], the exchange current of the redox reaction at the boundary of separation must depend on the stationary redox potential of the solution E_{ro} and be proportional to the concentration of the reagent C_o or C_r:

$$\left.\begin{aligned} i_0^n &= AC_r \exp\left\{\frac{-[\zeta^2\lambda + E_n + e(E_{ro} - E_0)]^2}{4\lambda\zeta^2 kT}\right\}, \\ i_0^p &= AC_o \exp\left\{\frac{-[\zeta^2\lambda + E_p - e(E_{ro} - E_0)]^2}{4\lambda\zeta^2 kT}\right\}. \end{aligned}\right\} \quad (6)$$

In these equations, i_o^n and i_o^p are the exchange currents of any redox reaction taking place at the electrode, A is a constant depending on the rate of movement along the reaction coordinates, C_o and C_r are the concentrations of the oxidizing and reducing agents, respectively, at the external plane of the Helmholtz part of the double layer at the equilibrium potential, and ζ is the number of electrons taking part in the reaction. The parameter λ characterizes the energy of transsolvation of the ion when the latter loses or acquires an electron as a result of the reaction [44]. An analysis of equation (4) leads to the conclusion that the limiting current i_0 must depend on the concentration of reducing agent or oxidizing agent:

$$i_0 \sim C_o(C_r). \quad (7)$$

After some rearrangements of equation (6) [42], it is also possible to show that

$$\frac{\partial \log i_0}{\partial E_s} \sim \frac{1}{120 \text{ mV}}. \quad (8)$$

The latter relation characterizes the dependence of the limiting current on the stationary redox potential of the solution. The exchange current can be observed experimentally, since at some applied voltage the current will not change with a rise in the voltage [38, 42]:

$$\left.\begin{aligned} i_n &= i_0^n\left(\frac{n}{n_0} - 1\right), \\ i_p &= i_0^p\left(1 - \frac{p}{p_0}\right), \end{aligned}\right\} \quad (9)$$

where i_n and i_p are the electronic and hole currents and n and p are the concentrations of electrons and holes at the contact in any section of the volt–ampere curve. The assumption that all the externally applied voltage falls in the insulator and the equilibrium at the contact with the electrolyte is not appreciably disturbed

TABLE 15. Hole Exchange Currents for Various Redox Systems at an Anthracene Electrode [39]

Redox system (0.1 M)	Inert electrolyte	Stationary potential (rel. NHE)	Exchange current A/cm^2
I^0/I^-	1 *M* KI	+0.54	$5\cdot10^{-8}$
$IrCl^{2-}/IrCl_6^{3-}$	0.5 *M* HCl	+0.96	$5\cdot10^{-6}$
$Cr_2O_7^{2-}/Cr^{3+}$	2 *M* H_2SO_4	+1.10	$5\cdot10^{-7}$
Mn^{3+}/Mn^{2+}	7.5 *M* H_2SO_4	+1.5	$1\cdot10^{-4}$
MnO_4^-/Mn^{2+}	0.5 *M* H_2SO_4	+1.5	$5\cdot10^{-5}$
Ce^{4+}/Ce^{3+}	2 *M* H_2SO_4	+1.44	$2\cdot10^{-5}$
Cl^{4+}/Cl^{3+}	7.5 *M* H_2SO_4	+1.40	$1\cdot10^{-4}$
Fe^{3+}/Fe^{2+}	0.5 *M* H_2SO_4	+0.77	10^{-11}
$Fe(CN)_6^{3-}/Fe(CN)_6^{4-}$	1 *M*	+0.69	11^{-11}

makes it possible to observe the exchange current of various redox systems experimentally. In addition, the equations given were obtained on the assumption that the surface states of a different type are unimportant. At the present time, voluminous material exists on exchange currents for various redox systems (Table 15). It is interesting to note that the exchange currents observed in anthracene are comparable with the currents flowing in a platinum electrode in the same solution.

REFERENCES

1. H. Kallmann and M. Pope, Nature (London), 185:753 (1960).
2. F. F. Vol'kenshtein, The Electronic Theory of Catalysis on Semiconductors [in Russian], Fizmatgiz, Moscow (1960).
3. K. Hauffe, Reaktionen in und an festen Stoffen, Springer, Berlin (1955).
4. R. S. Mulliken, J. Am. Chem. Soc., 74:811 (1952).
5. M. M. Labes, O. N. Rudyj, and P. L. Kronic, J. Am. Chem. Soc., 84:499 (1962).
6. T. C. Waddington and W. G. Schneider, Can. J. Chem., 36:789 (1958).
7. I. M. Assor and S. E. Harrison, J. Phys. Chem., 68:872 (1964).
8. G. H. Heilmeier and S. E. Harrison, Phys. Rev., 132:2010 (1963).
9. P. J. Rucroft, O. N. Rudyj, and M. M. Labes, J. Am. Chem. Soc., 85:2055 (1963).
10. M. M. Labes and O. N. Rudyj, J. Am. Chem. Soc., 85:2059 (1963).
11. H. Kallmann and M. Pope, J. Chem. Phys., 32:3000 (1960).
12. K. Kawasaki, K. Kanou, and M. Jizuka, Surface Science, 5:263 (1966).
13. A. Epstein and B. S. Wildi, J. Chem. Phys., 32:324 (1960).
14. E. I. Balabanov, E. L. Frankevich, and L. G. Cherkashina, Vysokomolek. Soedin., 5:1084 (1963).

15. V. S. Myl'nikov, Dokl. Akad. Nauk SSSR, 164:622 (1965).
16. G. H. Heilmeier and L. A. Zanoni, J. Phys. Chem. Solids, 25:603 (1964).
17. A. V. Vannikov, L. I. Boguslavskii, and V. B. Margulis, Fizika i Tekhnika Poluprovodnikov, 1:935 (1967).
18. L. I. Boguslavskii and V. B. Margulis, Proceedings of a Conference on Organic Semiconductors [in Russian], Izd. Akad. Nauk LatvSSR, Riga (1968).
19. V. B. Margulis and L. I. Boguslavskii, Élektrokhimiya, 3:329 (1967).
20. V. B. Margulis and L. I. Boguslavskii, Kinetika i Kataliz, 9:211 (1967).
21. I. A. Akimov, in: Elementary Photoprocesses in Molecules, A. N. Terenin (ed.) [in Russian], "Nauka," Moscow (1966), p. 397.
22. R. C. Nelson, J. Phys. Chem., 69:714 (1965).
23. H. Meier, J. Phys. Chem., 69:719 (1965).
24. B. J. Mulder and J. de Jonge, Koninkl. Ned. Acad. Wetenschap., B66:303 (1963).
25. I. Steketee and J. de Jonge, Koninkl. Ned. Acad. Wetenschap., B66:76 (1963).
26. V. S. Myl'nikov and A. N. Terenin, Dokl. Akad. Nauk SSSR, 155:1167 (1964).
27. W. Mehl and N. E. Wolf, J. Phys. Chem. Solids, 25:1221 (1964).
28. É. G. Sharoyan, N. N. Tikhomirova, and L. A. Blyumenfel'd, Zh. Strukt. Khim., 6:843 (1965).
29. M. Calvin, E. G. Cockbain, and M. Polanyi, Trans. Faraday Soc., 32:1436 (1936).
30. H. Miyagawa, W. Gordy, N. Watable, and H. M. Wilbur, Proc. Nat. Acad. Sci., 44:613 (1958).
31. D. D. Eley and S. Shooter, J. Catalysis, 2:259 (1963).
32. J. Uebersfeld, Ann. Phys., 1:395 (1956).
33. N. A. Bakh, V. D. Bityukov, A. V. Vannikov, and A. D. Grishina, Dokl. Akad. Nauk SSSR, 144:135 (1962).
34. D. E. Austen, Nature (London), 174:797 (1954).
35. N. N. Tikhomirova, B. V. Lukin, L. O. Razumova, and V. V. Voevodskii, Dokl. Akad. Nauk SSSR, 122:262 (1958).
36. S. Mrozowski and D. Wobshall, J. Chim. Phys., 57:915 (1960).
37. A. D. Grishina and A. V. Vannikov, Dokl. Akad. Nauk SSSR, 156:642 (1964).
38. A. D. Grishina, Zh. Strukt. Khim., 6:198 (1965).
39. W. Mehl, J. M. Hale, and F. Lohmann, Electrode Processes Symposium, Cleveland, Ohio, 1966.
40. M. Pope and H. Kallmann, J. Chem. Phys., 36:2486 (1962).
41. W. Mehl, Ber. Bunsenges Physik Chem., 69:583 (1965).
42. W. Mehl, J. M. Hale, and F. Lohmann, J. Electrochem. Soc., 113:1166 (1966).
43. P. Delahay (ed.), Advances in Electrochemistry and Electrochemical Engineering, Interscience, New York, Vol. 6 (1967), p. 399.
44. V. A. Myamlin and Yu. V. Pleskov, The Electrochemistry of Semiconductors [in Russian], "Nauka," Moscow (1965).
45. R. R. Dogonadze and Yu. A. Chizmadzhev, Dokl. Akad. Nauk SSSR, 145:563 (1962).
46. R. R. Dogonadze, A. M. Kuznetsov, and Yu. A. Chizmadzhev, Zh. Fiz. Khim., 38:652 (1964).
47. R. R. Dogonadze and Yu. A. Chizmadzhev, Dokl. Akad. Nauk SSSR, 144:463 (1962).

CHAPTER VI

ORGANIC SEMICONDUCTORS AS CATALYSTS

At the present time, numerous polymers have been synthesized which are stable at high temperatures and possess very diverse electrical and magnetic properties. Their catalytic activity in heterogeneous catalysis has been studied for the cases of the decomposition of hydrogen peroxide [1] and formic acid [2, 3], dehydrogenation and dehydration reactions [3, 5, 6], and the decomposition of hydrazine and nitric oxide [6, 8].

In the study of the catalytic activity of such systems, three main problems arise: in the first case, to find a correlation between the catalytic activity and the chemical structure of the polymer, which may throw light on the phenomenon of heterogeneous catalysis as a whole; in the second place, to obtain information on the structure of the polymer and its behavior during the occurrence of a chemical reaction at its surface; and, in the third place, to discover methods for the synthesis of specific organic catalysts which could be used in industry as effectively as in biochemical systems. Consequently, the study of organic semiconductors as catalysts is very attractive from the point of view of modelling enzymes, as well [7].

Desiring to gain an understanding of the laws directing such a complex biochemical system as the living organism, investigators assume that it is possible to study these laws on less complex subjects. Attempts to reproduce the functions of enzymes may not only open up one of the secrets of the living cell but also give the key to the creation of highly effective catalysts of very diverse chemical reactions.

CATALYTIC PROPERTIES OF THE PHTHALOCYANINES

The first studies of catalytic properties were undertaken by Polanyi et al. in 1936 [8]; they showed that the phthalocyanine ring is capable of activating hydrogen in the reaction of hydrogen compounds with oxygen.

As is well known, phthalocyanines are very similar in structural type to the porphyrin complex, the heme of the respiratory enzyme. Consequently, the interest of scientists in the results of a comparison of the catalytic properties of complexes of both classes is understandable. The catalase properties of the phthalocyanines of various metals have been studied [9]. The decomposition of hydrogen peroxide is convenient because its mechanism has been studied comparatively and at the same time it is one of the simplest enzymatic reactions. Complexes of bivalent iron possessed the greatest activity with respect to the decomposition of hydrogen peroxide. Because of the poor solubility of the phthalocyanines in water, 75% pyridine or, in some cases, concentrated sulfuric acid, was used a solvent.

Iron phthalocyanines promote the oxidation of unsaturated aliphatic compounds by atmospheric oxygen. In the oxidation reaction, the iron phthalocyanines function as oxygen carriers. The activating effect of a high-molecular-weight carrier on the catalase properties of iron phthalocyanine has been studied [9]. When the phthalocyanine was adsorbed on wood charcoal, its catalytic activity rose, just as in the case of the porphyrins. It is interesting that a catalyst adsorbed on charcoal readily detoxifies cyanides. Thus, to some extent it was possible to approach the model of an enzyme including a high-molecular-weight carrier, which is an integral part of any enzyme.

At the present time it has been established that enzymes consist of a protein part which apparently plays the role of carrier and heme — the organometallic part of the enzyme. In heme the metal is attached to the atoms involved in the chelate grouping by coordination bonds. In the opinion of Cook [9], the activity of the phthalocyanines that he observed is determined wholly by the central metal atom and not by the organic structure surrounding it. Consequently, the activity has a completely different nature from the catalytic activity of the phthalocyanine ring observed in the combination of hydrogen with oxygen by Polanyi et al. [8]. However, in later investigations it was shown that the catalytic activity does not depend only on the central atom included in the chelate.

CATALYTIC PROPERTIES OF CHELATE POLYMERS

Interesting prospects are opened up in the field of the study of the catalytic properties of chelate polymers. Claw-like (or chelate) complexes are generally insoluble in water and possess a high stability to the action of acids, alkalies, and reagents capable of reacting with the metal ion. Many chelate complexes are extremely stable to heat (400°C and above) [10]. Substances capable of forming intracomplex chelate compounds usually contain strongly polar electron-donating and electron-accepting groups. Most frequently, in the formation of the chelate grouping these groups are present in a chain of conugation which facilitates the migration of unshared electrons in the direction of the most electronegative atom or group.

A whole series of polychelates of the general formula

$$-\left[R-L\langle{}^{X}_{Y}\rangle Me \langle{}^{Y}_{X}\rangle L \right]_n -R-L,$$

where X and Y are strongly polarized electron-donating groups or atoms participating in the chelate grouping and connected with the metal by a donor–acceptor bond and L is a group of atoms connecting the chelate grouping with the radical R.

The catalytic activity of polychelates of copper, nickel, cobalt, iron, zinc, cadmium, manganese, and palladium with respect to the decomposition of hydrogen peroxide and hydrazine has been studied by Keier et al. [14-16]. The decomposition of hydrazine may take place in two directions: far-reaching decomposition to nitrogen and hydrogen

$$N_2H_4 \rightarrow N_2 + 2H_2; \tag{1}$$

and decomposition into ammonia and nitrogen

$$3\,N_2H_4 \rightarrow 4\,NH_3 + N_2. \tag{2}$$

Very frequently the process takes place simultaneously in both directions.

This feature of the reaction enabled the influence of the metal present in the chelate, the structure, and the chemical composition of the polymer on the selectivity of the catalyst to be established. The selectivity can be evaluated with respect to the rate of decomposition of hydrazine in the two directions shown. The reaction can take place wholly by equation (1) or (2) according to the chemical composition of the chelate node and the structure of the organic part of the polymer.

The investigation of the catalytic activity of various polychelates has shown that the overall rate of decomposition of hydrazine depends strongly both on the nature of the metal and on the nature of the atoms connected to the metal in the chelate node. On the basis of the decomposition of hydrazine it has been shown that the catalytic activity and the selectivity of polychelates are determined mainly by the nature of the metal present in the complex. Only polymers containing atoms of the transition metals (copper, nickel, cobalt) possess catalytic activity. Polychelates containing copper and nickel proved to be the most active. Their activity was two orders of magnitude greater than the activity of the sulfides of the same metals [17]. The second most important factor is the nature of the atoms connected to the metal in the chelate node. The third factor upon which the rate of decomposition of hydrazine depends is the nature of the organic radicals present in the polymer but not directly connected with the metal in the chelate grouping.

The result obtained in a comparison of monomeric analogs of a number of polychelates with the polymers deserve particular attention [15]. It was found that the monomers are considerably less active. This observation perhaps shows the possible role of collective effects in catalysts [18].

Apart from the phthalocyanines already mentioned, some other polymeric organic semiconductors have also proved to be good catalysts for the decomposition of hydrogen peroxide. A high catalytic activity has been observed in thermally treated polyacrylonitrile [10], polyaminoquinoline [20], and some polychelates [16]. The polychelates have been studied in most detail and we shall dwell on them more particularly. Just as in the decomposition of hydrazine, polychelates of the transition metals proved to be the most active. The nature of the addends substantially changes the catalytic activity of a polymer. The composition of the organic radical, even though it is not directly connected to the metal, can produce a change in the activity of the catalyst of three or four orders of magnitude. A study of the influence of the pH of the medium on the rate of decomposition of the peroxide showed that alkalinity led to an acceleration of the catalytic process. With a rise in the pH there is an increase in the degree of dissociation of the hydrogen peroxide molecule:

$$H_2O_2 \rightleftarrows H^+ + HOO^-, \qquad (3)$$

since the concentration of hydrogen ions decreases. The rate of the catalytic process changes in proportion to the change in the concentration of hydroxyl ions $[OH]^{1/4}$. The acceleration of the catalytic process with a rise in the pH shows the participation of peroxide ions in the reaction.

CATALYTIC PROPERTIES OF THERMALLY TREATED POLYACRYLONITRILE

In view of the known parallelism between the capacity for adopting the oxygen potential and the activity with respect to the decomposition of hydrogen peroxide, we must mention the behavior of thermally treated polyacrylonitrile as an oxygen electrode [21]. In alkaline solutions in an oxygen medium the stationary potential is only 25-30 mV more negative than that of the reversible oxygen electrode. With the course of time, the activity of the electrode decreases, changing by 90 mV after 1 h and by 120 mV after 30 h. It is obvious that the establishment of a reversible oxygen potential presupposes the reversibility of all the stages of ionization of oxygen and, in the first place, the stage of the adsorption of the O_2

molecule. Thus, the electrochemical behavior of the materials investigated must depend not only on their capacity for catalyzing the transition from hydrogen peroxide to water but also on their capacity for forming on the surface labile peroxides which readily split off oxygen, which is in fact characteristic of many polymers with conjugated bonds.

Thermally treated polyacrylonitrile has also proved to be extremely active in the dehydrogenation of formic acid [10]. The reaction was carried out at 240-300°C. One of the two samples investigated contained 0.01% of CuCl. The catalytic activity of both samples was not inferior to that of metallic and inorganic semiconducting catalysts for this reaction. The activation energy was determined from the dependence of the logarithm of the rate of the decomposition of the acid on the reciprocal temperature and was 21 kcal/mole for the sample containing copper and 25 kcal/mole for the other sample.

Pyrolyzed polyacrylonitrile, just like polyacetylene, is extremely active in the dehydrogenation of olefins, no saturated hydrocarbons or gaseous hydrogen being formed in the dehydrogenation process [22]. Such behavior is not characteristic of other carbonized materials and palladium, which catalyze the disproportionation reaction, or of industrial catalysts, the action of which is accompanied by the liberation of gaseous hydrogen.

Returning to the mechanism of the reaction it may be assumed that pyrolyzed polyacrylonitrile is a good hydrogen acceptor. The hydrogen is retained by chemical bonds until it is eliminated on the regeneration of the catalyst by treatment in an atmosphere of oxygen under very mild conditions. Such behavior of the catalyst can be best explained if we consider a structure of condensed pyridine rings:

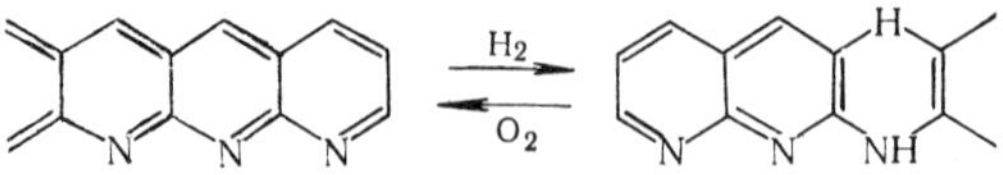

N N N $\underset{O_2}{\overset{H_2}{\rightleftharpoons}}$ N N NH H

Polyacrylonitrile also causes the migration of the double bond and cis-trans isomerization in olefinic systems. The mechanism of the transfer of hydrogen from the substrate to the surface of the catalyst has been studied by means of model compounds. It was found that hydrogen exists partly in the form of atomic hydrogen and partly in the form of hydride ion. A comparison of pyrolyzed polyacrylonitrile and the cyanoacetylene ion, which is structurally similar to it, has shown that they are considerably more active than other carbonized materials and behave specifically with respect to olefinic hydrocarbons.

The polymers obtained from β-chlorovinyl ketone [23] proved to be active catalysts for the oxidation of toluene with air at 370-380°C. Benzoic acid and benzaldehyde were isolated from the condensate. The specific activity of the samples was three orders of magnitude higher than that of carbon. The role of the radical states arising in organic semiconductors must be particularly stressed.

COMPARISON OF THE CATALYTIC PROPERTIES OF MONOMERIC AND POLYMERIC PHTHALOCYANINES

Acress and Eley [24] have studied the catalytic reactions

$$p\text{-}H_2 \rightarrow o\text{-}H_2$$

and

$$H_2 + D_2 \rightarrow 2HD$$

on copper phthalocyanine and two poly(copper phthalocyanine)s with different molecular weights. The results of the catalytic reaction were compared with the EPR spectra. On the basis of the results of the ortho-para transition of hydrogen it is possible to isolate three types of surface states participating in the catalytic reaction: 1) chelate copper atoms each with one unpaired d-electron, which give a broad EPR signal; 2) unpaired electrons formed by the rupture of C–C and C–H bonds on the thermal treatment of the polymers, which give a narrow EPR signal with a g-factor of ~2 (the intensity of the signal rises with an increase in the molecular weight; monomeric phthalocyanine does not give such a signal); 3) unpaired π-electrons formed by the cleavage of a double bond thermally or by reaction with H_2 or D_2:

$$D_2 + (\text{—CH=CH—}) \rightleftarrows (\text{—CHD·CHD—})$$

and

$$D_2 + (\text{—HC=N—}) \rightleftarrows (\text{—CHD·ND—}).$$

This process of creating unpaired π-electrons is equivalent to the formation of the biradical in a conjugated system:

$$(\text{—CH=CH—}) \rightarrow (\text{—}\overset{|}{\text{C}}\text{H—}\overset{|}{\text{C}}\text{H—}).$$

Like the process of the generation of an electron and a hole on intrinsic conduction in a semiconductor, the process of the origin of unpaired π-electrons is determined by ΔE, the thermal activation energy of conduction. As is well known, ΔE falls with an increase in the degree of conjugation. At temperatures of 373-523°K, the rate of the reaction $H_2 + D_2$ in polyphthalocyanine with the greater molecular weight was higher, which leads to the assumption that the role of the states determined by the π-electrons is fairly large (third type of states). The activation energy of the reaction was 10 kcal/mole.

At 200-373°K both the reactions studied took place with zero activation energy on polymeric phthalocyanines, which resembles the results obtained on films of the transition metals Fe, Co, and Ni. The absence of a catalytic effect for monomeric copper phthalocyanine unambiguously shows the role of the σ-electrons due to a certain number of cleaved C–C and C–H bonds (second type of states).

At temperatures from 27 to 200°K, the ortho–para transition of hydrogen took place with an activation energy of from 0 to 2 kcal/mole, which shows a paramagnetic interaction between the H_2 molecules present in the van der Waals layer in the paramagnetic state. It is possible that at these temperatures two types of state are active — σ-electrons and copper atoms.

CONNECTION BETWEEN PARAMAGNETISM AND CATALYTIC ACTIVITY

Apparently, polymers with a system of conjugation always contain local paramagnetic centers. In the opinion of a number of authors [25-27] the narrow EPR signals observed in such polymers are due to the presence of local paramagnetic centers arising as a consequence of a disturbance of the structure of the main chains of the macromolecules. The presence of paramagnetic centers in polymers with conjugated bonds has a pronounced effect on the chemical and electrical properties of the polymers.

This phenomenon has been studied in several investigations [27-32] and has acquired the name of the "local activation effect." The paramagnetic centers in polymers with a system of conjugation catalytically accelerate the polymerization of γ-chloropyridine and the block copolymerization of polyphenylacetylene, polyphenylene, and polyazophenylene with p-diethynylbenzene [23, 33]. The transfer of hydrogen from polynuclear aromatic compounds to stable free radicals is also catalyzed in the presence of paramagnetic centers [34, 35]. On the basis of these experimental data, several laws characteristic of the local activation effect have been deduced [36].

1. A paramagnetic polymer will exhibit the greatest activity if its complex with a diamagnetic substance forms a true or colloidal solution in the reaction medium.

2. The local paramagnetic centers displayed exert an activating influence on the reactivity of compounds with conjugated double bonds provided that the structures of the polymer containing the paramagnetic center and of the substance being activated are similar. In the contrary case, there is no activation process. Thus, for example, paramagnetic polyanthracene activates the inhibition by anthracene of the thermal oxidation of paraffins. But the addition of paramagnetic polyphenylene does not activate anthracene.

3. An increase in the polarity of the medium enhances the influence of the paramagnetic centers. For example, an investigation of the inhibiting activity of polyphenylene, polyphenylanthracene, anthracene, and ceresin on the oxidation of diethyl, dibutyl, and dioctyl phthalates, which contain different amounts of ester polar groups and consequently have different dielectric constants, showed that the polarity of the substrate undergoing oxidation affects the activity of the inhibitor.

4. In the presence of paramagnetic centers the observed activation energy of the process falls, although an increase in the

concentration of paramagnetic centers accelerates the process without decreasing the activation energy as a consequence of an increase in the pre-exponential factor.

In order to explain the local activation effect, it has been suggested that the local magnetic fields cause a disturbance of the electron clouds of the system. Generally, a chemical process takes place through excited singlet states, although the latter are higher than excited triplet states. Since a change in the multiplicity of the system is extremely unlikely, the spin—orbital interaction constants are small. However, if the system comes into the local magnetic field created by a paramagnetic center, the magnetic field acting on the reactants undergoes such sharp changes that the probability of a transition into the triplet state rises. An increase in the number of paramagnetic centers merely increases the number of elementary acts without changing the activation energy of the process.

In particular, on the basis of the experimental results obtained in a study of the mechanism of inhibition, it has been shown that the mechanism of the activating influence of the paramagnetic particles, which apparently can have an ion-radical nature, is determined by the polarization of the diamagnetic molecules of the inhibitor. The latter form complexes with the paramagnetic particles. Under these conditions, the transition of the inhibitor into the biradical triplet state is facilitated, which explains the increase in its efficiency as an inhibitor of redox processes. Information exists that the local paramagnetic centers may be not only responsible for the catalytic properties of some polymers but also closely connected with other physical properties of these substances.

Thus, for example, the optical and photoelectric properties of anthracene can be changed by doping carefully purified anthracene with the products of its pyrolysis with a mean molecular weight of ~1000. This doping fraction, containing 2.6×10^{18} spins/g with a g factor of 2.00, is obtained by the pyrolysis of anthracene at 450°C with subsequent chromatography on alumina. The fraction apparently consists of a polymer and possesses a developed system of conjugation, as can be judged from the absorption spectra taken in the ultraviolet and visible regions. The observation of the intrinsic fluorescence of anthracene as a function of the concentration of the paramagnetic particles and also the investigation of the pulse photoconductivity on a taumeter has shown that symbatic changes in the fluorescence and photocurrent yields take place with a change in the concentration of the paramagnetic centers. This is possibly connected with the fact that the radiationless decay of the excitons and the capture of the free carriers take place by the same mechanism.

Thus, in fact the question is that of the influence of local magnetic fields caused by paramagnetic particles on the chemical and physical properties of conjugated systems. The accumulated data permit the hope that in the not-too-distant future we shall have a clearer and better-founded picture of the action of the paramagnetic centers.

POSSIBLE APPROACHES TO AN EXPLANATION OF THE CATALYTIC ACTIVITY OF ORGANIC SEMICONDUCTORS

Even in the initial stage of investigations among organic semiconductors, catalysts for the decomposition of hydrogen peroxide, for oxidation, and for dehydrogenation were found. There is no doubt that further work in this field will considerably increase their number. Just as in the case of inorganic semiconductor catalysts, it is possible to attempt to connect the catalytic activity with the electrical conductivity of the polymers. From the point of view of electronic theory, the adsorbed molecules are present on the surface in the radical or valence-saturated state. The particles perform repeated transitions between these two states.

Thus, the catalytic activity is determined by the reactivity of the readsorbed molecules which, in its turn, is determined by the probability of the residence of the particles in the radical state. The catalytic activity of a surface depends on the spectrum of the local levels of all the molecules participating in the reaction and on the lowering of the Fermi level. This enables us to compare the catalytic activity with the work function and with the electrical conductivity. However, in the comparison of different catalysts the spectrum of the surface rarely remains unchanged. The mobility of the current carriers also usually changes, particularly if the measurements are carried out on powders. Consequently, very frequently a comparison of these characteristics does not give the desired result [37]. In actual fact, a comparison carried out for polychelates has shown that there is no direct correlation between the bulk electrical properties and the catalytic activity. It would be more correct to seek a correlation between the catalytic properties, the concentration of electrons, and the width of the forbidden gap. However, a comparison of the electrical conductivity of various polychelates can hardly be successful since this also depends on the mobility of the current carriers, which may differ extremely widely in different substances. Consequently, in this case, as in the study of the mechanism of conduction, the difficulty consists in investigating the mobility and concentration of the carriers in organic semiconductors.

It will be impossible in general to compare the catalytic and electrical properties (as only polymers containing transition metals are active) and to interpret the results by means of the theory of the crystal field [25]. In this case, adsorption and catalysis are determined mainly by the electronic state of the metal in the chelate node. The results of EPR measurements on the adsorption of hydrogen on polyphthalocyanine are said to confirm this point of view [24] (cf. Chapter V).

However, the influence of the substrate is not excluded, its action being shown by the change in the fine electronic structure of the chelately-bound metal. This influence has been followed from the x-ray absorption spectra of copper complexly bound with aliphatic and aromatic radicals [39]. The results obtained show that polymers with aliphatic radicals have a lower (in comparison with aromatic radicals) intensity of transitions to the 4p level and indicate a change in the electronic state of the chelately bound metal.

The study of the hyperfine structure of the EPR spectra of catalysts in the substitution of hydrogen in the phenyl rings of an organic radical also indicates the influence of changes in remote orders in the electronic state of the chelate atom. This means that through the π-electrons the organic part of the molecule takes part in the formation of the bond between the copper ions, the nitrogen, and the sulfur in the chelate grouping. This scheme makes it possible to assume that the unpaired electrons are also delocalized on the ligands, while a superexchange between the different parts of the polymer is possible through them. Thanks to the presence of superexchange, the polymer differs from the monomer since in the case of the polymer the copper participates in the formation of the whole chain of conjugation through the 3d-orbitals. It is possible that the catalytic activity of polymers is due to a lowering of the outer filled level of the copper (d_z^2) which is determined by the degree of the nonplanar π-bond. Such a change in the EPR spectrum correlates with the change in the catalytic activity with respect to the decomposition of a hydroperoxide [37].

However, from all these points of view it is difficult to understand why polymeric chelates are, as a rule, more active than monomers of the same composition; the electronic state of the metal in the chelate grouping is the same in both cases. What has been said above, and also the activating influence of the substrate on the catalytic properties of the phthalocyanines, forces us to assume that the molecule must be considered as a whole.

In a comparison of organic and inorganic semiconductors one feature was noted [18] which may be important in a consideration both of organic semiconductors and of biological materials in general. This feature is that the electron is considerably more localized in organic semiconductors. Then we have to deal with semiconductors with narrow bands, and for electronic exchange to be possible (in the case of a redox reaction, for example) the electronic levels of the reacting molecules must be within the limits of the narrow band of the organic semiconductor. It is not excluded that the hypothesis put forward may prove useful for understanding the specificity of reactions in such systems. Thus, the theory of the crystal field and the ideas of the band theory are the starting points of all attempts to explain the catalytic activity of organic semiconductors.

CATALYTIC ACTIVITY OF BIOPOLYMERS

A completely different approach is being developed in a consideration of the enzymatic activity of biopolymers. The influence of the prosthetic group on the catalytic activity has been established definitely with any type of catalysis. However, the role of the protein prosthetic group and whether or not it participated

TABLE 16. Change in the Entropies and Energies of Activation of Some Enzymatic Reactions in Various Temperature Regions [44]

Enzyme	Substrate	ΔE, kcal	$\Delta(\Delta S)$, units	β/T_{mean}
Invertase	Sucrose	45	186	1.05
Trypsin	Casein	50	182	1.02
Lipase	Tributyrin	30	180	—
Urease	Urea	4	12	1.1
Fumarase	Fumarate	4	12	1.1
Catalase	Hydrogen peroxide	25	87	0.95
Luciferase	Luciferin	18	—	—
Amylase	Starch	8	—	—

directly in the structural and energy transformations remained unclear. Koshland's hypothesis [40-43] connects the catalytic activity with the metallochemical transformations in the biopolymers.

Let us consider the simplest scheme of enzymatic transformation, which may be described by the equation

$$E + S \underset{k_2}{\overset{k_1}{\rightleftarrows}} ES \overset{k_3}{\rightarrow} E + P, \tag{4}$$

where E is the enzyme, ES is the enzyme–substrate complex, S is the substrate, and P the reaction product. By experiment we can find only the combination of the constants k_1 and k_3.

$$K_m = \frac{k_1}{k_1 + k_3}, \tag{5}$$

where K_m is the Michaelis constant.

Then the entropy and energy of activation are calculated from the equation

$$K_m = \frac{kT}{h} \exp\frac{\Delta S}{R} \exp\left(-\frac{\Delta E}{RT}\right). \tag{6}$$

The multiplicity of the different types of reactions has been confirmed by Likhtenshtein [44]. It was shown that it is possible to construct empirical relationships for the activation energy ΔE and the enthalpy of the reaction H [3]:

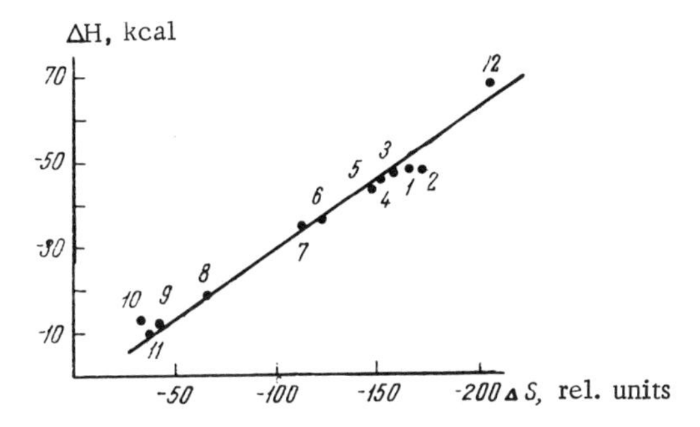

Fig. 38. Dependence of the heats on the entropies of the inhibition of the luminescence of the bacteria *A. tisheri:* 1) chlorothion; 2) trianol; 3) barbital; 4) evinol; 5) phenobarbital; 6) chloral hydrate; 7) amytal; 8) salicylamide; 9) benzamide; 10) sulfanilamide on the bacterium *V. phosphorescens;* 11) sulfanilamide; 12) urethane on the bacterium *V. phosphoreum.*

$$\Delta E = A_1 + \beta_1 \Delta S, \quad H = A_2 + \beta_2 \Delta S, \tag{7}$$

where A_1, A_2, β_1, and β_2 are the coefficients which are constant for a given series of reactions. These relationships are preserved over extremely wide ranges (20 kcal/mole for ΔE) (Table 16).

The pronounced change in the activation energy should lead to a change in the rate of the reaction by 10–15 orders of magnitude, but this does not take place because of the change in entropy. A similar compensation effect has been considered for very diverse catalytic reactions [46]. Figure 38 shows the relationship between the heats and entropies of the inhibition of the luminescense of bacteria. Equations (7) are valid for extremely voluminous experimental material obtained on very diverse systems the only common property of which is the presence of protein molecules. According to the ideas developed by Sukhorukov and Likhtenshtein [45], denaturation is a "melting," a disordering of the native structure of the biopolymer. It is possible that enzymatic reactions in biopolymers are connected with the same physical causes. In order to understand this assumption we must assume that in the performance of some elementary act in catalysis at an active center a definite rearrangement of the adjacent part of the protein molecule or of water molecules adjacent to the active center of the enzyme must take place. Then the stages of the catalytic process can be described in the following way [44].

The catalytic process begins with the enzyme and the substrate acquiring as a consequence of fluctuation sufficient energy to overcome the activation barrier of the reaction. The reaction liberates a certain amount of energy, and part of it goes to "melting" in sections adjacent to the enzyme and part is dissipated in the form of heat. Thus, spontaneous rhythmic conformations of part of the protein molecule take place. Because of this, part of the energy liberated as a result of the chemical reaction is restored to the contact groups of the enzyme. This hypothesis has not yet been confirmed experimentally and time will show how far it corresponds to reality. For us, however, this approach is interesting because it stresses the role of conformational changes in the prosthetic group during the action of the enzyme.

REFERENCES

1. V. A. Kargin, A. V. Topchiev, B. A. Krentsel', L. S. Polak, and B. É. Davydov, Zh. Vses. Khim. Obshchestva im D. I. Mendeleeva, 5:507 (1960).
2. E. S. Dokukina, S. Z. Roginskii, and M. M. Sakharov, Dokl. Akad. Nauk SSSR, 137:893, 1404 (1961).
3. N. P. Keier and I. U. Astaf'ev, Kinetika i Kataliz, 3:364 (1962).
4. S. Z. Roginskii, A. A. Berlin, and L. M. Kittseva, Kinetika i Kataliz, 4:118 (1963).
5. J. Gallard, M. Nechtschein, M. Soutif, et al., Bull. Soc. Chim., 1963:2204.
6. J. Gallard, Th. Laederlich, M. Nechtschein, et al., Proceedings of the Third Conference of Heterogeneous Catalysis, Amsterdam, 1964.
7. L. A. Nikolaev, Usp. Khim., 33:580 (1965).
8. M. Calvin, E. G. Cockbain, and M. Polanyi, Trans. Faraday Soc., 32:1436 (1936).
9. A. H. Cook, J. Chem. Soc., 1938:1761, 1768, 1774.
10. A. A. Berlin and N. G. Matveeva, Usp. Khim., 29:277 (1960).
11. A. P. Terent'ev, V. V. Rodé, and E. G. Rukhadze, Vysokomolek. Soedin., 2:1557 (1960).
12. A. P. Terent'ev, E. G. Rukhadze, N. G. Mochalina, and G. V. Panova, Heterochain High-Molecular-Weight Compounds [in Russian], "Nauka," Moscow (1964), p. 123.
13. A. P. Terent'ev, I. G. Mochalina, E. G. Rukhadze, and E. M. Povolotskaya, Vysokomolek. Soedin., 6:1267 (1964).
14. N. P. Keier, G. K. Boreskov, V. V. Rodé, A. P. Terent'ev, and E. G. Rukhadze, Kinetika i Kataliz, 2:509 (1961).

15. N. P. Keier, G. K. Boreskov, L. F. Rubtsova, and E. G. Rukhadze, Kinetika i Kataliz, 3:691 (1962).
16. N. P. Keier, M. T. Troitskaya, and E. G. Rukhadze, Kinetika i Kataliz, 3:691 (1962).
17. G. K. Boreskov, N. P. Keier, L. F. Rubtsova, and E. G. Rukhadze, Dokl. Akad. Nauk SSSR, 144:1069 (1962).
18. M. G. Evans and J. Gersely, Biochim. et Biophys. Acta, 3:188 (1949).
19. A. V. Topchiev, M. A. Geiderikh, B. É. Davydov, V. A. Kargin, B. A. Krentsel', I. M. Kustanovich, and L. S. Polak, Dokl. Akad. Nauk SSSR, 128:312 (1959).
20. A. A. Berlin, L. A. Blyumenfel'd, and N. N. Semenov, Izv. Akad. Nauk SSSR, Ser. Khim., 1959:1689.
21. A. N. Frumkin, L. I. Boguslavskii, and V. S. Serebrennikov, Dokl. Akad. Nauk SSSR, 142:878 (1962).
22. J. Manassen and J. Wallach, J. Am. Chem. Soc., 87:2671 (1965).
23. A. N. Nesmeyanov, A. M. Rubinshtein, A. A. Dulov, A. A. Slinkin, M. I. Rybinskaya, and G. A. Slonimskii, Dokl. Akad. Nauk SSSR, 135:609 (1960).
24. G. J. Acress and D. D. Eley, Trans. Faraday Soc., 60:1157 (1964).
25. A. A. Berlin, Khim. Prom., 1962(12):23.
26. L. A. Blyumenfel'd and V. A. Benderskii, Zh. Strukt. Khim., 4:405 (1963).
27. A. A. Berlin, V. A. Bonsyatskii, and L. A. Lyubchenko, Izv. Akad. Nauk SSSR, Otd. Khim. Nauk, 1962:1312.
28. A. A. Berlin and S. I. Bass, Izv. Akad. Nauk SSSR, Otd. Khim. Nauk, 1962:1492.
29. A. A. Berlin and S. I. Bass, Teoret. Éksperim. Khim., Akad. Nauk Ukr. SSR, 1:151 (1965).
30. A. A. Berlin, L. A. Blyumenfel'd, M. I. Cherkashin, A. É. Kalmanson, and O. G. Sel'skaya, Vysokomolek. Soedin., 1:1361 (1961).
31. A. A. Berlin, V. A. Grigorovskaya, V. P. Parini, and Kh. Gafurov, Dokl. Akad. Nauk SSSR, 156:1370 (1964).
32. A. A. Berlin and V. A. Bonsyatskii, Dokl. Akad. Nauk SSSR, 154:627 (1964).
33. A. A. Berlin and S. I. Bass, Dokl. Akad. Nauk SSSR, 150:795 (1963).
34. A. A. Berlin, V. A. Bonsyatskii, and B. I. Liogon'kii, Dokl. Akad. Nauk SSSR, 144:1316 (1962).
35. A. A. Berlin and L. A. Blyumenfel'd, Izv. Akad. Nauk SSSR, Otd. Khim. Nauk, 1964:1720.
36. A. A. Berlin and S. I. Bass, Izv. Akad. Nauk SSSR, Otd. Khim. Nauk, 1963:1854.
37. Collection: Problems of Kinetics and Catalysis [in Russian], "Nauka," Moscow (1966), p. 66.
38. C. Ballhausen, Introduction to Ligand Field Theory, McGraw-Hill, New York (1962).
39. R. P. Akopdzhanov, E. E. Vainshtein, N. P. Keier, L. M. Kefeli, and E. G. Rukhadze, Kinetika i Kataliz, 5:616 (1964).
40. A. E. Braunshtein (ed.), Enzymes [in Russian], "Nauka," Moscow (1964).
41. M. V. Vol'kenshtein, Molecules and Life, Plenum Press, New York (1970).
42. D. E. Koshland, in: The Enzymes, P. D. Boyer, H. Lardy, and K. Myrbäck (eds.), Academic Press, New York, Vol. I (1959), p. 305.

43. K. Linderstorm-Land and J. Shellman, in: The Enzymes, P. D. Boyer, H. Lardy, and K. Myrbäck (eds.), Academic Press, New York, Vol. I (1959), p. 443.
44. G. I. Likhtenshtein, Biofizika, 11:24 (1966).
45. B. I. Sukhorukov and G. I. Likhtenshtein, Biofizika, 10:933 (1965).
46. Yu. L. Khait and S. Z. Roginskii, Dokl. Akad. Nauk SSSR, 130:366 (1960).

CHAPTER VII

BIOLOGY AND ORGANIC SEMICONDUCTORS*

The question of the possibility of applying to biopolymers the ideas developed in the physics of the solid state has a fairly long history. In 1938 Jordan [1] and in 1941 Szent-Györgyi [2] suggested that proteins may possess the nature of semiconductors. Subsequent biophysical investigations of biopolymers at the molecular level led to the conviction that the phenomena taking place in ordered biological structures cannot be interpreted unless in some cases the biopolymer is considered as a solid body. This is explained, in the first place, by the fact that proteins consist of very large molecules with molecular weights reckoned in millions and, in the second place, by the fact that the enormous role played by the medium (for example, a solvent) in which the various processes take place is becoming ever clearer to chemists and biologists.

The main biochemical processes in living organisms, which are directly connected with the transfer of energy and of electrons and protons, cannot be explained within the framework of classical physics. A deficiency of the ideas of a purely chemical mechanism is also found on considering the mechanism of photosynthesis, vision, and the transmission of hereditary information. The same applies to bioluminescence, which is observed in a whole series of organisms. The chemical reactions in bioluminescence are ac-

* In the present chapter, in addition to the main material on the electrical conductivity of biopolymers the authors give in the shortest possible form some elementary information on the structure of the mitochondria and chloroplasts in order to show more clearly the role and position of electrical phenomena in the occurrence of many important biological processes.

companied by the emission of photons. The processes of bioluminescence are interesting because with their aid it is possible to observe electronic excitations and the conversion of chemical energy into the energy of motion of electrons on irradiation. In redox processes taking place with the participation of the cytochrome system the change in the valence of the chelately bound iron ion must also take place with the transfer of an electron. However, as a rule the phenomena mentioned are included in a chain of very diverse transformations which greatly complicates their study. As an example we may mention the processes of energy transformation taking place in the mitochondria.

TRANSFORMATION OF ENERGY AND TRANSFER OF ELECTRONS IN BIOLOGICAL SYSTEMS

In 1949, it was established that the site of the generation of adenosine triphosphate — the main source of energy in the cells of the living organism — is formed by the very fine granules called "mitochondria" [3].

Electron-microscope studies enabled many basic features of the structure of the mitochondrion and its membrane to be established. Figure 39 shows part of the membrane and a general reconstruction of a mitochondrion. The interior of the mitochondrion is surrounded by two membranes of which the outer one is smooth while the inner has long projections — the cristae. Because of these features of their structure, the mitochondria have a fairly large surface. The inner membrane with the cristae contains enzyme complexes by means of which the energy of the nutritional substances penetrating into the cell is converted into the energy of the macroergic bonds of adenosine triphosphate (ATP). However, before we go in more detail into the mechanism of the various reactions taking place in the mitochondria we must consider the question of the transformation of energy in metabolism.

In any organism there is an enormous number of different substances reacting with one another. They form a complex network of conjugated chemical reactions. The direction and rate of these reactions is determined by a series of regulating conditions

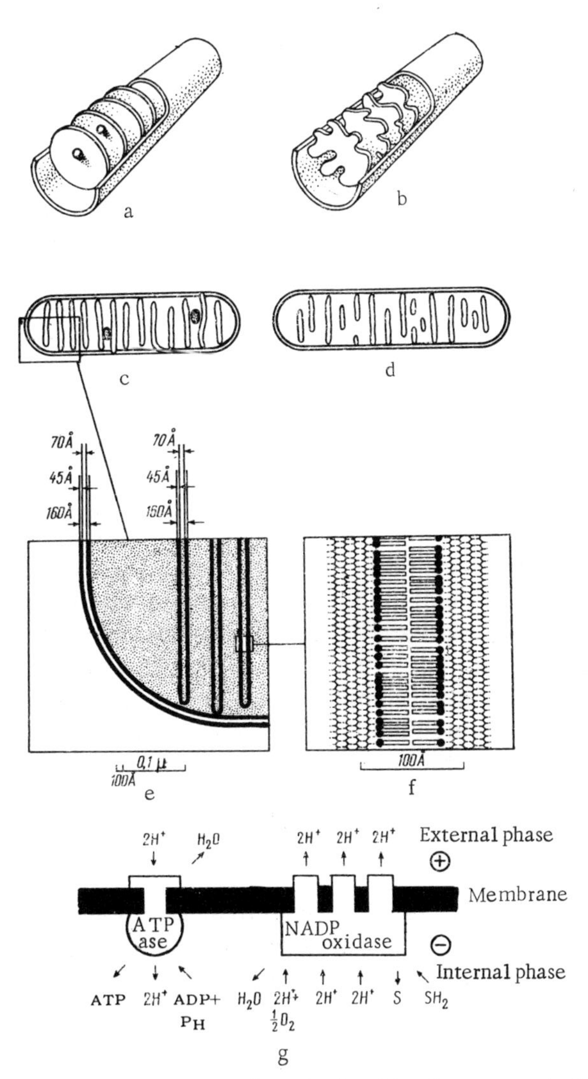

Fig. 39. Reconstruction of a mitochondrion from an analysis of electron-micrographs. a, b) Three-dimensional; c, d) on a plane; e) structure of the inner and outer membranes at high magnification; f) structure of the inner membrane alone (in the center is a bimolecular phospholipid layer and at the edges, in the form of cells, layers of protein with various enzymes); g) model to explain the conjugation of the oxidation and phosphorylation reactions [6].

among which the most important are thermodynamic criteria. From this point of view, the series of chemical reactions is only a source of energy ensuring the occurrence of the various processes in the living organism. The living cell requires energy for its existence. The gradual decomposition of the nutritional substances serves as "fuel" for the cell. The combustion of the nutritional substances takes place through a series of stages and is accompanied by the absorption of oxygen and the liberation of carbon dioxide and water. This exchange of gases has acquired the name of "respiration." The study of this phenomenon has led to the necessity for investigating the mechanisms of the liberation of energy on respiration. It has been found that a whole system of successive oxidation–reduction reactions take place which form the so-called respiratory chain in which a number of enzymes take part:

Dehydrogenases → flavoproteins → cytochromes

Conjugated with respiration is the process of phosphorylation — the formation of ATP, the most important energy carrier, which is formed in the cell during the process of metabolism. In addition to the cells in which various forms of useful energy are obtained through the decomposition of organic substances, microorganisms exist which are capable of oxidizing inorganic substrates. However, the cells that contain chlorophyll use light energy. Three main types of transformations exist in which ATP is the final product. These are glycolysis, oxidative phosphorylation, and photosynthetic phosphorylation [4].

The effectiveness of the chemical reactions serving as the source of energy in the living organism can be expressed by measuring the free energy at constant pressure (the Gibbs function):

$$\Delta Z = \Delta H - T\Delta S, \tag{1}$$

where ΔH and ΔS are the change in enthalpy and entropy, respectively, as a result of the chemical reaction.

The process of oxidation is a multistage process. Three groups of enzymes take part in it. It is very convenient to discuss the energy relationships in the transfer of electrons from the substance undergoing oxidation to the oxidizing agent by means of redox potential. As is well known, in a reversible redox system

$$A + e \rightleftarrows B \tag{2}$$

an equilibrium exists, and the redox potential of such a system is defined by the relation

$$\varphi = \varphi_0 + \frac{RT}{nF} \log \frac{A}{B}. \tag{3}$$

In a state of cquilibrium under standard conditions

$$\Delta Z = -RT \ln K. \tag{4}$$

The same equation can be expressed in electrical units:

$$\Delta\varphi = \frac{RT}{nF} \ln K, \tag{5}$$

where K is the equilibrium constant.

Consequently, measurements of the redox potential in a cell, in its individual parts, or in the isolated enzymes, give valuable information on the metabolism and permit the sequence of the arrangement of the various enzymes in the respiratory chain according to the redox potentials to be determined. This permits a schematic representation of the metabolic process in the form of a chain of conjugated redox reactions. At the top, positive, end of the chain is oxygen and at the bottom hydrogen. Between the substrate subjected to oxidation and the oxygen are located a number of intermediate substances. The conversion of one of the members of this series causes successive reactions through the chain to its positive end in the direction toward the oxygen.

The interaction of the substrate with the dehydrogenases leads to the dehydrogenation of the substrate, i.e., to the simultaneous transfer of a proton and an electron along the respiratory chain. The overall reaction can be written

$$AH_2 + B \rightarrow A + BH_2,$$

where AH_2 is the substrate and B is the acceptor, which may be one of the coenzymes: nicotinamide adenine dinucleotide (NAD) and flavin adenine nucleotide (FAD).

Two classes of dehydrogenases are known [4]: anaerobic enzymes, which transfer hydrogen to various acceptors but not to oxygen, and aerobic dehydrogenases which transfer hydrogen to molecular oxygen. NAD and NADP are coenzymes of the anaerobic dehydrogenases. Each substrate has its own enzyme. The enzymes differ from one another by the protein moiety but have the same coenzyme (for example, NAD and NADP). Dehydrogenation, consisting in the transfer of hydrogen to NAD, takes place in the protein moiety of the enzyme. Since NAD is soluble, between the various dehydrogenases a reversible equilibrium with respect to NAD exists which permits the smooth transfer of hydrogen along the chain of enzymatic reactions to take place. This equilibrium may be likened to some extent to that which exists for polybasic acids with different smoothly changing dissociation constants.

The pyridine enzymes play an important part in processes of alcoholic and lactic fermentation. However, oxidation is generally not limited to this stage and takes place in a group of flavin enzymes and then in the cytochrome system which also forms the essence of aerobic oxidation. The flavin enzymes have a more positive redox potential. An equilibrium exists between the flavin enzymes and the soluble coenzyme flavin adenine nucleotide or flavin mononucleotide.

The flavin enzymes transfer hydrogen from the pyridine nucleotides to the cytochromic system. The latter is also an enzyme chain which transfers hydrogen indirectly to atmospheric oxygen. However, before entering the cytochromic system the hydrogen enters coenzyme Q, which is an intermediary between the flavoproteins and the cytochromes. It is interesting that one part of the enzyme has a hydrocarbon tail, which is soluble in lipids, and another contains quinone derivatives, i.e., it is polar.

The cytochromes consist of claw-like compounds having as the central chelate atom iron or copper the valence of which can change reversibly during the process of transferring an electron along the respiratory chain. Hydrogen is not transferred by a cytochromic system. Electrons from the cytochromes become attached to oxygen:

$$O_2 + e \rightarrow O_2^-.$$

Then the O_2^- reacts with the protons from the solutions that arise in the oxidation of the substrate. We may observe that in the respiratory chain a two-electron transfer takes place first and then, along the cytochrome system, a one-electron transfer.

In actual fact, in the reaction of the substrate with NAD the transfer of one electron and one hydrogen atom takes place, i.e., two equivalents. In the subsequent stage, reaction with a flavoprotein, the transfer of two reduced equivalents again takes place. However, the action of the reduced flavine with cytochrome-b and the subsequent transfer of charge is a one-electron process. Consequently, in addition to scheme (6) another scheme is possible in which a flavoprotein reacts with two cytochromes simultaneously, (7).

$$\text{Substrate} \rightleftarrows \text{NAD} \rightleftarrows \text{FP} \rightarrow b \rightarrow c \rightarrow c_1 \rightarrow a \rightarrow a_3 \rightarrow O_2. \tag{6}$$

$$\text{Substrate} \rightleftarrows \text{NAD} \rightleftarrows \text{FP} \begin{array}{l} \nearrow b \rightarrow c \rightarrow c_1 \rightarrow a \rightarrow a_3 \searrow \\ \searrow b \rightarrow c \rightarrow c_1 \rightarrow a \rightarrow a_3 \nearrow \end{array} O_2. \tag{7}$$

If the substrate undergoing oxidation is a hexose, the overall reaction has the following form:

$$C_6H_{12}O_6 + 6\,O_2 \rightarrow 6\,H_2O + 6\,CO_2 + 686 \text{ kcal/mole.}$$

As can be seen from the equation, in addition to water, carbon dioxide is liberated, this being produced not by the direct oxidation of carbon but as a result of the decarboxylation of organic acids by means of carboxylases.

Yet another result of the oxidation reaction is the liberation of energy. Let us now consider how the liberated energy is transformed on the formation of macroergic bonds by the aid of which all types of useful work in the organism are performed.

The phosphorylation reaction can be written in the form of the equation

$$BH_2 + A + ATP \rightleftarrows B + AH_2 + ADP + P_H.$$

For this there is the equilibrium

$$K = \frac{[ADP]\,[P_H]}{[ATP]},$$

TABLE 17. Change in the Redox Potential of Electron Carriers and Points of Oxidative Phosphorylation [3]

Electron carriers	$\Delta\varphi_0$, V	Points of phosphorylation	
		found	calculated
NAD–flavoprotein	0.27	+	+
Flavoprotein–cytochrome-b	0.09	0	0
Cytochrome-b–cytochrome-c	0.22	+	+
Cytochrome-c–cytochrome-a	0.03	0	0
Cytochrome-a–oxygen	0.53	+	+

provided that the concentration of the other components is unchanged. The same relation can be expressed on the scale of redox potentials according to equation (5). Thus, we have a complex system which is reversible with respect to ATP, NAD, a proton, and an electron. Consequently, in the oxidation process the interconversion of a redox potential and a "phosphoryl" potential in fact takes place.

In order to elucidate at what points of the respiratory chain phosphorylation is possible it is necessary to direct our attention to the energy interrelationships in the respiratory chain. The "value" of the macroergic ATP bond is 9 kcal/mole, which corresponds precisely to the free energy in the transfer of a pair of electrons in a circuit with a potential difference of 0.20 V. Table 17 shows the change in the redox potentials of electron carriers in the respiratory chain. As follows from the table, fairly large changes in free energy exist only in the segments denoted by plus signs. These thermodynamic calculations have been confirmed experimentally. The agreement of the results obtained by thermodynamic calculation and by experiment is all the more convincing since the thermodynamic estimates are not accurate because they were deduced for the case of a closed system in equilibrium while the respiratory chain is an open system in a state of stationary equilibrium.

The overall scheme of oxidative phosphorylation was put forward by Lehninger [3], who suggested that each stage of the respiratory chain conjugated with phosphorylation separates a nonconjugated stage from the subsequent conjugated one.

For phosphorylation to take place in conjugation with the oxidative process, the two reactions must have a common intermediate product. It is just in this case that the transfer of chemical energy from a donor reaction to an acceptor reaction becomes possible. A series of substances exists which disturb the conjugation of the oxidation and phosphorylation reactions. These substances, so-called uncoupling agents, enable us to draw some conclusions concerning the mechanism of the conjugated reactions. Dinitrophenol is among the most fully studied uncoupling agents. Although it possesses an uncoupling action, it does not suppress respiration. A possible mechanism of the action of dinitrophenol will be discussed below. However, at the moment we shall turn our attention to the exchange by H^+ and OH^- ions in the process of oxidative phosphorylation. In the transfer of electrons along the respiratory chain, the dehydrogenation of the substrate with the aid of NAD forms NAD-H_2. In the reduction of flavoproteins by NAD-H_2, H^+ is bound with the formation of FAD-H_2.

When an electron with a reduced flavoprotein passes through coenzyme Q to cytochrome-b, two protons are again released. In the reduction of oxygen by cytochrome-a and -a_3, hydroxyl ions are generated. Although the overall equation does not lead to the formation of an excess of H^+ or OH^- ions, the points of formation of these ions in the respiratory chain are disconnected. In the synthesis of ATP at pH ~7, H^+ ion is bound in accordance with the reaction

$$ADP^{-3} + HPO_4^{-2} + H^+ \rightleftarrows ATP^{-4} + H_2O.$$

Evidently, in the very structure of the mitochondria the formation and binding of H^+ and OH^- ions may act as the most important keys regulating the ratio of rates of respiration and phosphorylation and may be the most important elements of the processes of the active transfer of ions.

According to Mitchell's scheme [5, 6] the dehydrogenation of the substrate causes the liberation of a protein, which appears on the outer side of the membrane (Fig. 39g). The reduction of oxygen is accompanied by the liberation of OH^- ions and the disappearance of a molecule of water at the lower surface. This is possible only where the membrane is impermeable to H^+ and OH^- and when enzymes with specific orientation are present on its surface. The H^+ and OH^- ions accumulate on different sides of the membrane. Thus, the oxidative process leads to the appearance of a difference in the potentials of the two sides of the membrane and may be the cause of the active transfer of ions. Finally, the accumulation of H^+ and OH^- ions switches on the action of the enzyme ATPase, which leads to the elimination of the H^+ and OH^- ions and the synthesis of ATP. Mitchell's hypothesis apparently comprises one of the most essential features of the process of conjugation — its symmetry — thereby emphasizing the role of the membrane as a physicochemical factor.

Thus, the essential difference of Mitchell's scheme from the chemical mechanisms of oxidative phosphorylation is that here conjugation appears not in the existence of two intermediate products but in the difference between the electrochemical potentials existing at the membrane.

According to this scheme, uncoupling agents of the type of dinitrophenol, which increase the permeability of the membranes for H^+ ions, lead to a disturbance of the potential gradient and to the cessation of the phosphorylation process, as is observed experimentally.

It would be extremely attractive to compare Mitchell's scheme with the actual structure of the membrane. Consequently, we shall consider in more detail the ultrastructure of the membrane since the latter not only acts as an envelope for the mitochondria but also fulfills extremely diverse functions of active transfer and contains highly organized associations of respiratory enzymes.

Basically, the membrane consists of phospholipids and structural protein, which form the structure of the membrane (see Fig. 39). The lipids form a stable membrane which, together with the structural protein, is the skeleton for the other elements of the membrane.

The proteins present at the inner and outer surfaces are probably not bound by means of strong covalent bonds with the phospholipids of the membrane. This may be deduced from the fact that the former are easily extracted with a mixture of chloroform and methanol. The proteins are fixed to the elements of the membrane by the electrostatic interaction of certain protein and phospholipid groups. Such interaction leads to the formation of

TABLE 18. Components of the Respiratory Association [3]

Electron-transferring enzyme	Molecular weight
NAD · H_2 dehydrogenase	1,000,000
Succinate dehydrogenase	200,000
Cytochrome-b	28,000
Cytochrome-c_1	40,000
Cytochrome-c	12,000
Monomeric cytochrome a	70,000
Total	1,350,000

stable complexes of the phospholipid with the protein. For example, a complex of cytochrome-c with a phospholipid has been isolated which has been called "lipocytochrome." A protein has also been isolated from mitochondria which apparently has structural functions. By weight it composes 55% of the total mass of the protein present in the mitochondrial membranes. The structural protein forms a complex with the phospholipids of the membrane and, in this way, acts as a link between the functional elements and the structural elements. It is interesting that the cytochromes, phospholipids, and structural protein tend to polymerization with the formation of extremely stable associations exhibiting a lamellar structure.

The double lipid layer in the membrane is ~60 Å thick. The coefficient of permeability of natural membranes for various nonelectrolytes is proportional to the partition coefficient between olive oil and water. Biological membranes are characterized by very high electrical resistance and capacity. The respiratory associations consisting of a complete set of electron carriers and conjugated enzymes amounts to 25% of the total mass of the protein of the membranes. Calculation has shown that there are 17,000 molecules of cytochrome-a to one liver cell mitochondrion. However, the number of cytochromes in each association is unknown. The respiratory associations are localized mainly in the cristae. Table 18 gives the components of the respiratory association.

If, however, to these components are added the enzyme strongly bound to the association, the weight of this maximum association will be 1,830,000. Figure 39g gives the plan of the arrangement of the respiratory association in the mitochondrial membrane.

The chemistry of oxidative phosphorylation reactions is extremely complex and each investigation yields more and more new elements which make the general scheme ever more cumbersome. However, the mechanism of oxidative phosphorylation can hardly be finally understood merely as a combination of various chemical reactions without taking into account the role of the mitochondrial membrane itself, in which the enzymes are packed in such a way that the transfer of the reacting molecules in a definite direction becomes possible. At the same time, this permits the role of the lipid-protein layer to be explained precisely as a phase in which important stages of the oxidative phosphorylation process take place. In studying the membrane as a phase, it is important to elucidate how the transfer of ions in this phase takes place: whether the transfer of electrons in the membrane is possible and what is the mechanism of conduction in it. Also obscure is just how the transfer of an electron in the respiratory chain takes place.

The electrons are used to reduce the Fe^{3+} in the heme of the first cytochrome. This heme is similar in structure to the well-studied copper phthalocyanine. Even if it is assumed that the coenzyme can react directly with the heme of the first cytochrome, the question unavoidably arises as to how the first cytochrome reacts with the second, the second with the third, and so on [7]. Moreover, as shown by Erenberg and Theorell [9] the heme is "hidden" in the wrinkles of the protein structure and is not present directly on the surface. The interaction of the hemes with the neighboring cytochromes represents a problem which can be solved from the point of view of the motion of the electrons in protein structures.

CONSIDERATION OF BIOPOLYMERS WITHIN THE FRAMEWORK OF THE BAND MODEL

A band model for proteins was first constructed on the basis of the sequence of bonds $-C=O \cdots H-N$ [10] in the form of an extended polypeptide chain. The width of the conduction band proved to be ~0.2 eV and the effective mass of the charge carriers considerably greater than the mass of a free electron. The width of the forbidden gap was calculated as 3 eV. Moreover, a calculation has been carried out for a polypeptide chain in the form of an α-helix [11]. In more recent papers, this value has been calculated as 5×3.95 eV [8, 9, 12, 13]. One paper [14] considers polypeptides as a one-dimensional system consisting of peptide groups linked by hydrogen bonds. The conduction band was obtained in the approximation of a strong bond and the exciton zone by the method of Frenkel' and Peierls [12, 13].

The width of the band for holes or electrons was found to be about 0.1 eV. For a singlet exciton it proved to be 0.5 eV and for a triplet exciton 10^{-5} eV. The results given show that the interaction of electrons with vibration plays an important part in the mechanism of the transfer of charge through polypeptides since the frequency of the vibrations of the peptide blocks, with the exception of the frequencies of the vibrations of the N–H bond, which is 0.4 eV/$\hbar$, is in the region of 0.2 eV/$\hbar$. However, the frequency of the intramolecular N–H← O→ vibrations is ~0.01 eV/$\hbar$. The physical significance of the interaction of electrons with vibrations can be understood from the following discussion.

Let us assume that an electron or hole in a polypeptide is localized on any peptide group. Then, if the life of this state is comparable with the period of interpeptide vibrations the distances between all the bonds in the peptide group are changed and stabilize this state. Furthermore, in the neighborhood of this peptide group the distances between the neighboring peptides also become different, which changes the probability of the transfer from group to group. We may observe that the proposed mechanism is extremely similar to the mechanism of the motion of a polaron in oxide semiconductors.

It has been established [15] that protons possess hole conductivity with an extremely high activation energy (2.5–3 eV). However, appreciable conductivity in proteins has been detected only at temperatures so high that they are found in no living organism. In the majority of cases, ionic reagents are complexly bound to proteins or to nucleic acids and can give charge-transfer complexes which favor the formation of holes.

Some of the facts indirectly show that the holes or electrons take part in biological processes. On the irradiation of proteins, as the EPR signal shows, in strictly definite positions (adjacent to cysteine) traps capturing holes are formed. This perhaps shows the presence of mobile holes in the individual molecules [16].

A complication in the study of biopolymers is that they contain large amounts of water. If before measurement the biopolymer is dried, many of its properties become completely different from those which it had in the living organism. In a study of protein substances containing water it is difficult to distinguish the electronic conductivity from the ionic conductivity. Consequently, the results of experimental studies carried out with different proteins are extremely contradictory. Some authors state that ionic [17] or protonic [18, 19] conduction exists in a protein. Others have come to the conclusion that proteins are electronic semiconductors [20-31].

A number of authors [32, 33] have studied the conductivity of DNA. Duchesne et al. [26] found that the activation energy of conduction calculated from the ratio E/2kT is 1.8 eV. According to Pauling et al. [11] the activation energy is 2.4 eV. This discrepancy in the values of the activation energy has been ascribed to the influence of adsorbed water. It was found that the electrical conductivity of the salt decreases after vacuum treatment for 12 days.

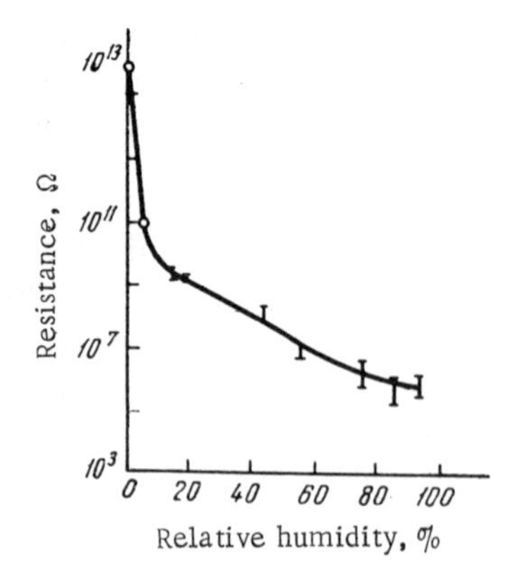

Fig. 40. Influence of the relative humidity on the resistance of the sodium salt of DNA (R).

TABLE 19. Change in the Electrical Characteristics of the Sodium Salt of DNA on Drying [28]

Time of vacuum treatment, days	R, Ω	Activation energy, eV	
		dark conductivity	photo-conductivity
1	$2 \cdot 10^{8}$	1.3	0.7
2	$5 \cdot 10^{9}$	1.83	0.8
5	$5 \cdot 10^{11}$	2.15	0.98
12	10^{14}	2.43	1.16

The action of adsorbed water vapor on the conductivity of the sodium salt of DNA has been studied in detail by Liang and Scala [28].

Figure 40 shows how the relative humidity affects the electrical conductivity of a film, the equilibrium value of the conductivity being established the more slowly the greater the relative humidity. Attempts to explain the influence of light on the conductivity of proteins have been made [29, 30]. It was found that the conductivity of films of gelatin increases tens of times on irradiation with ultraviolet light. The conductivity of blood albumen somewhat increases even on illumination with visible light. The photoconductivity of films of hemoglobin, gelatin, and gelatin sensitized with auramine has been studied by Nelson [31]. The photoconductivity rose on illumination with light having a wavelength shorter than 300 mμ, i.e., in the region of the intrinsic absorption of the protein. The role of the heme and the dye apparently reduced simply to increasing the absolute value of the photocurrent. In a study of the photoconductivity of the sodium salt of DNA at different humidities, it was observed [28] that the photocurrent of dried samples rises linearly with the applied voltage, and the activation energy of the photocurrent calculated from the ratio E/kT is 1.16 eV. The figure obtained is considerably higher than the values of the activation energy of the photocurrent characteristic of the other organic semiconductors. Table 19 shows how the resistance, the activation energy of the dark current, and the activation energy of the photocurrent change when the sodium salt of DNA is dried.

USE OF THE HETEROGENEOUS MODEL FOR EXPLAINING THE ELECTRICAL PROPERTIES OF BIOPOLYMERS

In a discussion of the results of a study of the electrical conductivity of biopolymers carried out with direct current, one must take into account both the intramolecular and intermolecular transfer of current carriers. Detailed measurements performed on polymeric organic semiconductors have shown that a comparatively high conductivity and low activation energy of intramolecular transfer do not appear in measurements with constant current (cf. Chapter IV).

In a study of the parameters of proteins and polyphosphates as functions of the frequency of alternating current, the dispersion of the electrical conductivity and of the activation energy with a change in the frequency was found [32]. This dispersion can be explained by the polydispersity of the sample, each macromolecule being represented as a microcrystal. The activation energy of the conduction of DNA at a frequency of 10^9 kHz was 0.2 eV.

The dispersion of the electrical conductivity with a change in the frequency has also been observed in solutions of DNA. This was explained by the rotational movements of the elongated DNA molecules [33]; however, as follows from the work of Pollak [34], the most successful explanation of this dispersion is achieved by using the Maxwell–Wagner model, according to which the dispersion must be ascribed to conduction along the DNA molecules.

The mobility of the carriers in many biopolymers has been calculated from measurements of the Hall effect at a frequency of 10^{10} Hz. The results obtained were explained by means of a model based on the assumption that each macromolecule can be regarded as a conducting sphere with a radius of 10^{-5}cm [35, 36]. As a whole, the system consists of N spheres chaotically distributed in an insulator. Since a solution of the problem was found for the polarization $\vec{p}$ of a sphere of a conducting nonmagnetic material present in a dielectric medium with ε_0, the specific polarization of the medium can be found as

$$\vec{P} = N\vec{p}.$$

By the definition of complex conductivity,

$$\vec{\sigma} = \frac{\vec{I}}{\vec{E}},$$

where the current density in the medium

$$\vec{I} = \frac{1}{4\pi} \cdot \frac{d\vec{D}}{dt},$$

and the induction is connected with the specific polarization by the relation

$$\vec{D} = \vec{E} + 4\pi\vec{P}.$$

By substituting in the equation the corresponding expression for the longitudinal and transverse polarization of the sphere and separating the real part from $\vec{\sigma}$ it is possible to find the longitudinal (ohmic) and transverse (Hall) components of the conductivity. The effective mobility can be determined by the equation

$$\mu = -\frac{c}{H_0}\sigma_{\perp}/\sigma_{\parallel},$$

where H_0 is the strength of the constant magnetic field. The model used may differ extremely widely from the actual structure of the sample since the conducting regions do not have the form of spheres, the specific volume of the conducting inclusions is not infinitely small, and the "infinite ideal dielectric" is neither infinite nor an ideal insulator and conducts the current. In addition to this, the Debye dipole losses, for example, when water is present in the native samples, may affect the results of measurements.

Nevertheless, the study of very different materials by this method can give an idea of the order of magnitude of the mobility and the sign of the current carriers. It is also clear that the results of Table 20 are comparable only to the extent to which the model selected is valid. Only a direct measurement of the mobility — for example, by the method described in Chapter IV — can give reliable values of the mobilities in biological materials.

An attempt has been made [47] to show the validity of the hypothesis that the membrane of mitochondria does not only serve as a "frame" for the most convenient mutual arrangement of the

TABLE 20. Electrical Characteristics of Some Biopolymers at a Frequency of 10^{10} Hz [32]

Substance	Experimental conditions	Water content, moles/100 (dry subst.)	ρ, $\Omega\cdot$cm	μ, cm²/sec	Type of carrier
Glycine	Dark	Normal	$1.4\cdot10^4$	1.65±85%	*n*
Diglycine	"	"		1.6±50%	*n*
Triglycine	"	"	$1.5\cdot10^3$	2±50%	*n*
Tetraglycine	"	"	$1\cdot10^3$	1.5±50%	*n*
Oligoleucine	"	"	$1.1\cdot10^4$	0	—
Polyalanine	"	0,12	$1\cdot10^3$	0	—
	"	None	$4.2\cdot10^3$	0	—
	Light	0.12	($\Delta\rho = 12\%$)	2.4±40%	*n*
	Dark	0.12		0.8±75%	*n*
Cytochrome-c	"	0.8	$0.7\cdot10^3$	0	—
	Light	None	$2.7\cdot10^3$	1±80%	*n*
	"	"	($\Delta\rho = 3{,}5\%$)	1.3±70%	*n*
Hemoglobin	Dark	0.8	$1.1\cdot10^3$	2±50%	*n*
	Light	0.8	($\Delta\rho = 30\%$)	3.2±40%	*n*
	Dark		$3\cdot10^3$	2±60%	*p*
Hemoglobin	"		$1.2\cdot10^3$	0	—
DNA	"	Normal	$1\cdot10^3$	0.5±40%	*p*
	"	"	($\Delta\rho = 3\%$)	0.85±50%	*p*
DNA (denatured)	"	None	$2\cdot10^4$	0	—

enzymes but also participates directly in the transfer of electrons. To study the processes of charge transfer in mitochondria, a suspension of the light fraction of the mitochondria isolated from rat liver was transferred in the form of a drop to a plate with conducting contacts. The sample was immediately placed in a vacuum chamber and pumped out until a compact film 10^{-2} cm thick had been formed. The upper electrode was vaporized on from aluminum, particular precautions being taken to prevent the heating of the sample. All the results were obtained by using the method described on p. 80.

In films completely deaerated in a high vacuum for several days, no continuous motion through the film of charges generated by pulses of low-energy electrons was observed. The time of movement of the charges was less than 10^{-6} sec. The amplitude of the measuring pulse increased linearly with a rise in the field strength.

In order to find the distance through which the charged particles passed in the electric field, it was necessary to estimate the number of charges formed per unit of absorbed energy G, which in its turn was determined by the ditribution of the secondary electrons with respect to energies. Since the form of this distribution for primary electrons with an energy of ~ 3 keV was unknown, it was possible to estimate G only roughly. On the assumption that the static dielectric constant ε is 5-7 and a contribution to the measured current is given by charges weakly connected by Coulomb interaction and escaping initial recombination, it is possible to estimate G as between 1.0 and 0.3 pairs of charges per 100 eV. In this case, in a field of 10^4 V/cm the charged particles travelled a distance of 10^{-7} - 3×10^{-7} cm, i.e., they did not leave the region of generation.

As a study of the volt–ampere characteristics in deaerated samples possessing a specific resistance of $10^{11}\ \Omega \cdot$cm showed, an electronic mechanism of conduction is apparently involved. This was confirmed by the absence of polarization.

The fact that the carriers travelled a distance of not more than 30 Å in the applied electric field indicated that there was no appreciable displacement of the charges under the action of the attractive field. Apparently in this case the carriers did not leave the limits of some structural unit.

The situation changed radically when the film contained water (in an amount not exceeding 1%). In this case the continuous movement of the carriers through the whole of the film was observed and it was possible to measure the mobility of the charge carriers. The considerable value of the mobility of the positive charges (2×10^{-2} cm^2/V·sec) obtained in these experiments permitted the assumption that the phenomenon involved was not ionic but electronic conduction. The mobility of protons in aqueous solutions at infinite dilution is 3.6×10^{-3} cm^2/V·sec. In the system studied the mobility of a proton could be only less than this figure if the formation of protonic bands was not assumed.

The mobility of the negative charges was estimated as $\sim 10^{-2}$ cm^2/V·sec, while the mobility of OH^- is 1.8×10^{-3} cm^2/V·sec. (It is interesting to observe that in polymers with semiconducting properties at the same conductivity as in the samples studied, the mobility of the current carriers does not exceed 10^{-4} $cm^2/V \cdot sec$, which is two orders of magnitude smaller.) The activation energy of the mobility was 0.15 eV. An increase in the mobility of the carriers with the temperature could be explained by the presence of traps or be connected with the existence of boundaries of separation.

In order to find whether the observed mobility was actually connected in some way with the structure of the biopolymers entering into the composition of the mitochondria, samples were heated to 125°C for 15 min. After heating, the observed signal due to the motion of the positive charges disappeared. Subsequent exposure to water vapor did not lead to a restoration of the initial mobility observed in the native sample. According to rough estimates, the displacement of the charges in the denatured film did not exceed 10-30 Å.

A confirmation of the electronic nature of the conduction was also the volt–ampere relationship for samples containing water. In this case, at low field strengths slight polarization was observed which disappeared completely when the strength of the field was increased.

In a study of the heavy fraction of mitochondria no continuous motion of the carriers through the specimen was observed. When the membranes were disturbed by the freezing and subsequent pumping out of the mitochondria continuous conduction appeared. However, the continuous motion of the carriers was observed only where the sample contained moisture (less than 1%).

In a study by this method of individual structural units — ultrasonic fragments of membranes of the mitochondria and a preparation of cytochrome-c individually — no continuous motion of the carriers was observed. However, in the case of the fragments the carriers moved in an electric field a distance 2-2.5 times greater than in the case of cytochrome-c and a dehydrated film of the light fraction of the mitochondria. The absence of continuous motion of the carriers in films of ultrasonic fragments of the membranes of mitochondria is possibly connected with the fact that the content of water in these preparations was inadequate because of the very rapid loss of water on vacuum treatment.

Thus a series of indications of the electronic mechanism of

conduction in the biological materials investigated may be listed: the stability of the conduction in a constant current in dry DNA; the fact that the conductivity is independent of the moisture content in the region of very high frequencies; the decrease in the high-frequency conductivity when the sample is denatured; the noncomparability of the high-frequency conductivity for different counter-ions; the photoconductivity effect; and the Hall effect at superhigh frequencies.

EXCITATION PROCESSES IN BIOLOGICAL MACROMOLECULES

In addition to the processes of conduction in biological macromolecules, processes of excitation are extremely important. Since proteins have an ordered structure, not all excited states are possible; in other words, the excited states form an exciton band.

The intensity of the absorption of light is determined by the probability of the transfer of a molecule from the ground state into an excited state and may serve as a characteristic of the orderedness of the structure. In polypeptides the dispersion of the optical rotation has been used as a powerful factor for determining their structure [37]. The transfer of excitations between the individual molecules of chlorophyll plays an essential role in the first stage of photosynthesis. The source of energy in the transfer of excitation may be irradiation or a chemical reaction. The main "problem" of a plant is to use the energy of the photons and to convert it into chemical energy. Some stages of this complex biochemical process have a purely chemical nature and others cannot be understood without making use of ideas developed in the physics of solids, in particular without using the concept of the migration of excitons. In order to establish the main routes of the photosynthetic reactions it was necessary to study the physical properties of the pigments. Extremely thin films of chlorophyll were applied to an electrode, and the photopotentials arising when the electrode was illuminated were observed. The experiments were carried out in connection with the "semiconductor" model of the action of chlorophyll in which it was assumed that on illumination the direct sepa-

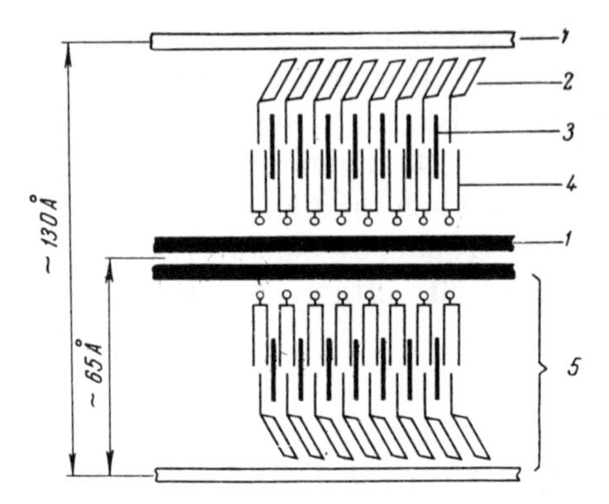

Fig. 41. Assumed mutual arrangement of the various structural units in a chloroplast. 1) Protein; 2) chlorophyll; 3) carotenoids; 4) phospholipids; 5) elementary membrane.

ration of an electron and a hole takes place in chlorophyll under the action of the light. The mobility of the holes in chlorophyll proved to be extremely small [38].

The migration of excitons takes place in structures known as chloroplasts (Fig. 41) [39]. The cell of a green leaf generally contains no more than a hundred chloroplasts which consist of plates 5-10 μ long and about 2 μ in diameter. The main substance of the chloroplasts is chlorophyll. Chlorophyll consists of a porphyrin nucleus with conjugated bonds, which is hydrophilic, and a fatty hydrophobic "tail."

Many modifications of the chlorophylls exist which differ in the substituents in the porphyrin ring. For example, chlorophyll-a is a reduced form and chlorophyll-b an oxidized form. These forms have different absorption spectra. Chlorophyll may exist both in the amorphous and in the crystalline state. If one judges from the absorption spectra, in chloroplasts the chlorophyll may have a different degree of aggregation. It possibly forms an amorphous monomolecular layer adjacent to the phospholipid layer.

The chloroplasts also contain carotenoids and phycoerythrins. The carotenes and their oxidized forms — xanthophylls — comprise systems of conjugated double bonds. In the chloroplast they are present between the phytol tails of the chlorophyll in the form of bundles 20-25 Å high, three molecules of carotenoids being associated with each molecule of chlorophyll. The phycoerythrins and their oxidized form — phycocyanins — are present in the protein layer and the ratio of the oxidized and reduced forms for all three types of pigments is determined by the redox potential of the medium. The lipids and the proteins amount to 35-50% of the dry weight of the chloroplasts. Very small amounts (one molecule per 300-400 molecules of chlorophyll) of hemoproteins, cytochrome-f, cytochrome-b_6, and quinones are also present. This shows the presence of electron-transfer chains. On excitation, chlorophyll can reversibly emit or add an electron.

On participating in the sensitization reaction, chlorophyll may undergo both oxidation and reduction [38]:

$$\mathrm{Chl} + h\nu \rightarrow \mathrm{Chl}^*,$$
$$\mathrm{Chl}^* + \mathrm{B} \rightarrow \mathrm{Chl}^+$$
$$\dot{\mathrm{C}}\mathrm{hl}^+ + \mathrm{RH} \rightarrow \mathrm{Chl} + \dot{\mathrm{R}}\mathrm{H}^+,$$

or

$$\mathrm{Chl} + h\nu \rightarrow \mathrm{Chl}^*,$$
$$\mathrm{Chl}^* + \mathrm{RH} \rightarrow \dot{\mathrm{C}}\mathrm{hl}^- + \dot{\mathrm{R}}\mathrm{H}^+,$$
$$\dot{\mathrm{C}}\mathrm{hl}^- + \mathrm{B} \rightarrow \mathrm{Chl} + \dot{\mathrm{B}}^-.$$

Apparently, the order of the partial reactions is not constant and depends on the state of the pigment and the oxidizing and reducing agents taking part in the reaction.

The energy of the photons that is adsorbed by the chlorophyll can be transferred by an exciton mechanism. However, it is not yet clear whether the separation of the charges occurs at the site where the holes or electrons are absorbed or whether the exciton disintegrates elsewhere and then the migration of the charges takes place. In any case, electronic paramagnetic resonance shows the appearance of free electrons in the process of photosynthesis [40, 41].

MOBILITY OF THE PROTONS IN THE DNA MOLECULE AND A POSSIBLE MECHANISM OF AGING

Watson and Crick (see [4]) have constructed a model of DNA in the form of a double helix which agrees well with the experimental data. The helices consist of carbohydrate-phosphate groups and are connected in pairs through purine or pyridine bases by means of hydrogen bonds. The following bases are involved in the composition of DNA: adenine (A), thymine (T), guanidine (G), and cytosine (C). These bases contain groups some of which are proton donors and others proton acceptors. Each pair of bases is connected by two or more H bonds which, as it were, fill up the interior of the helix. The formation of H bonds leads to spatial restrictions and therefore in the arrangement of the bases a certain regularity appears and only pairs of definite form exist, i.e., each base corresponds to its "complementary" base.

A proton proves to be shared between two electron pairs on oxygen or nitrogen atoms (Fig. 42) [42]. Thus, if a definite sequence of bases ATGACTG exists, then the same sequence of "complementary" bases but in the reverse order TACTGAC arises from the other end of the helix. This is the basic genetic code.

The DNA molecule can reproduce itself from the nucleotide material present in the surrounding medium. For a more detailed consideration of the formation of hydrogen bonds between the bases, Fig. 42 must be somewhat schematized:

$$A\left\{\begin{matrix} :\mathrm{H} \\ : \end{matrix}\right. \quad C\left\{\begin{matrix} :\mathrm{H} \\ : \\ : \end{matrix}\right. \quad \left.\begin{matrix} : \\ \mathrm{H}: \\ : \end{matrix}\right\}T \quad \left.\begin{matrix} : \\ \mathrm{H}: \\ \mathrm{H}: \end{matrix}\right\}G.$$

However, another tautomeric form is possible which can be denoted by A* C* T* and G*. Then instead of the pairs shown in the preceding scheme, the following new forms will be produced:

$$A^*\left\{\begin{matrix} : \\ :\mathrm{H} \end{matrix}\right. \quad C^*\left\{\begin{matrix} : \\ :\mathrm{H} \\ : \end{matrix}\right. \quad \left.\begin{matrix} \mathrm{H}: \\ : \\ : \end{matrix}\right\}T^* \quad \left.\begin{matrix} \mathrm{H}: \\ : \\ \mathrm{H}: \end{matrix}\right\}G^* \qquad (8)$$

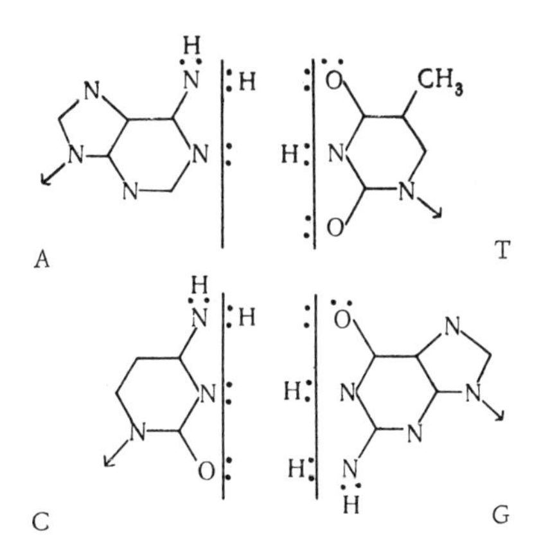

Fig. 42. Bonding of the nucleotide bases in one of the elements of the DNA double helix. A = adenine; T = thymine; C = cytosine; G = guanidine.

The new forms will not combine with one another in the same way as before: A–T, G–C, but differently: A*–C, A–C*, G*–T, G–T*. This means that the "complementary" bases have become different. Thus an error appears in the genetic information which can be explained by the following scheme:

Original sequence	A G T C A T T G C A.
Tautomeric form	A G T*C A T T G C A.
Complementary sequence	T C G G T A A C G T.
New sequence.............	A G C C A T T G C A.

The number of DNA molecules bearing erroneous information rises rapidly as a result of the synthesis of new molecules. Thus, there is a common cause of the wonderful stability of the hereditary material over thousands of years and of the disordered mutations of the genetic code which are caused by quantum jumps of a proton between various stationary states.

In order to understand the hypothesis developed by Löwdin [42] more deeply, let us dwell in somewhat more detail on the nature of the hydrogen bond [43].

By an intermolecular hydrogen bond is understood an association of similar or different molecules with definite functional groups. One of these groups must be a proton donor (acid group) and the other an electron donor (basic group).

Consequently, the proton is displaced in the direction of one of the molecules. If the molecules are similar as, for example, in the case of the pyridine bases, two positions of the equilibrium, which are equivalent, are possible. The proton jumps between these two positions. The energy scheme of this process can be represented in the form of two potential troughs separated by a barrier. In the case of the DNA molecule, it would be interesting to discuss the possibility of the conjugated exchange of protons, for example, in the adenine–thymine pair:

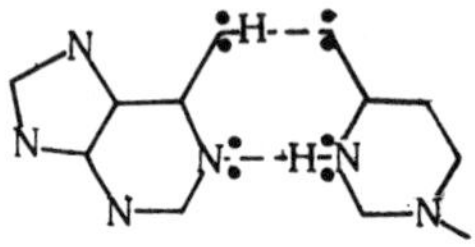

The transition of protons from one state to another is a quantum-mechanical problem and can be considered from the point of view of a tunnel transition. A similar but simpler problem has been considered in the case of proton conductivity [44]. In spite of all the attractiveness of this hypothesis it must be approached critically. Apparently, in a more correct model one must take into account the interaction of the proton with the medium in the same way as was first done by Marcus [45] and then more strictly in a consideration of the problem of the charge transfer of ions in solution [46].

The spontaneous change of the hydrogen bonds in the DNA molecule which was considered above leads to the situation that with time there is a loss of the genetic code, and this process may be regarded as ageing. The change in the code may lead to such a sharp change in the DNA that the extremely fine mechanisms controlling the rate of synthesis of new molecules may be disturbed.

REFERENCES

1. P. Jordan, Naturwiss., 26:693 (1938).
2. A. Szent-Györgyi, Nature (London), 148:158 (1941).
3. A. Lehninger, The Mitochondrion: Molecular Basis of Structure and Function, W. A. Benjamin, New York (1964).
4. A. G. Pasynskii, Biophysical Chemistry [in Russian], Vysshaya shkola," Moscow (1963).
5. P. Mitchell, Nature (London), 191:144 (1961).
6. P. Mitchell, Nature (London), 214:1327 (1967).
7. J. Gergely, Symposium on Electrical Conductivity in Organic Solids, H. Kallmann (ed.) (1960).

8. F. S. Sjostrand, Rev. Mod. Phys., 31:301 (1959).
9. A. Erenberg and H. Theorell, Acta Chem. Scand., 9:1193 (1955).
10. M. G. Evans and J. Gergely, Biochim. et Biophys. Acta, 3:188 (1949).
11. L. Pauling, R. B. Corey, and H. R. Branson, Proc. Nat. Acad. Sci. US, 37:205 (1951).
12. M. Suard, G. Berthver, and B. Pullman, Biochim. et Biophys. Acta, 52:954 (1961).
13. J. Ladik, Acta Physiol. Acad. Scient. Hung., 15:287 (1963).
14. H. Suziki, A. M. Minesaki, and S. Yomosa, J. Phys. Soc. Japan, 19:2175 (1964).
15. M. H. Cardew and D. D. Eley, Disc. Faraday Soc., 27:161 (1959).
16. M. Kotani, Rev. Mod. Phys., 35:717 (1963).
17. G. Kins and J. A. Medley, J. Colloid Sci., 4:1 (1949).
18. J. Baxter, Trans. Faraday Soc., 39:207 (1958).
19. N. V. Ril', Zh. Fiz. Khim., 29:1537 (1955).
20. S. Marčič, G. Pifat, and V. Pravdič, Biochim. et Biophys. Acta, 79:293 (1964).
21. A. T. Vartanyan, Dokl. Akad. Nauk SSSR, 143:1317 (1962).
22. B. Rosenberg, J. Chem. Phys., 36:816 (1962).
23. M. H. Cardew and D. D. Eley, Disc. Faraday Soc., 27:115 (1959).
24. D. D. Eley and D. J. Spivey, Trans. Faraday Soc., 56:1432 (1960).
25. Yu. A. Vladimirov and K. N. Timofeev, Biofizika, 11:33 (1966).
26. J. Duchesne, J. Depireux, A. Bertinchaps, N. Cornet, and J. M. van der Kaa, Nature (London), 188:405 (1960).
27. D. D. Eley and D. J. Spivey, Trans. Faraday Soc., 58:411 (1962).
28. C. Y. Liang and E. G. Scala, J. Chem. Phys., 40:919 (1964).
29. C. Y. Liang and E. G. Scala, Nature (London), 198:86 (1963).
30. P. Douzon and J. M. Thuillier, J. Chim. Phys. et Phys.-Chim., 57:96 (1960).
31. R. C. Nelson, J. Chem. Phys., 39:112 (1963).
32. C. T. O'Konsky, Rev. Mod. Phys., 35:721 (1963).
33. S. Takashima, J. Mol. Biol., 7:455 (1965).
34. M. Pollak, J. Chem. Phys., 43:908 (1965).
35. É. M. Trukhan, Biofizika, 11:412 (1965).
36. É. M. Trukhan, Fiz. Tverd. Tela, 4:3496 (1963).
37. W. Moffitt, J. Chem. Phys., 25:467 (1956).
38. The Biochemistry and Biophysics of Photosynthesis [in Russian], "Nauka," Moscow (1965).
39. M. Calvin, Rev. Mod. Phys., 31:331 (1959).
40. B. Commoner, J. J. Heise, B. B. Lippincott, R. E. Norbert, J. V. Passonneau, and T. Towsend, Science, 120:57 (1959).
41. M. Calvin, Rev. Mod. Phys., 31:157 (1959).
42. Per-Olov Löwdin, Rev. Mod. Phys., 35:724 (1963).
43. G. Pimentel and O. MacClennan, The Hydrogen Bond, W. H. Freeman, San Francisco (1960).
44. P. Taylor, Disc. Faraday Soc., 27:237 (1959).
45. R. A. Marcus, J. Phys. Chem., 67:853 (1963).
46. R. R. Dogonadze, A. M. Kuznetsov, and Yu. A. Chizmadzhev, Dokl. Akad. Nauk SSSR, 144:563 (1962).
47. A. V. Vannikov and L. I. Boguslavskii, Biofizika, Vol. 14, No. 3 (1969).

CHAPTER VIII

PROSPECTS OF THE PRACTICAL APPLICATION OF ORGANIC SEMICONDUCTORS

At the present time, hundreds of new organic semiconducting materials have been obtained [1-3]. In this chapter we shall consider those characteristics of these materials that can be used in various branches of power engineering and radioelectronics [4].

THE USE OF THE ELECTRICAL PROPERTIES OF ORGANIC SEMICONDUCTORS

First of all, let us dwell on the possibility of using materials combining semiconducting properties with the properties characteristics of polymers. Polymers have been obtained which have conductivities of from 10^{-13} to $10^{3}\,\Omega^{-1}\cdot\text{cm}^{-1}$ with positive and negative temperature coefficients of resistance, thermoplastic and thermosetting, with both p- and n-type carriers. The mobility of the charge carriers in these substances is between 10^{-8} and 100 $\text{cm}^2/\text{V}\cdot\text{sec}$ and the thermo-emf varies between 1000 and 3 μV/deg. In addition, not only diamagnetic and paramagnetic but also ferromagnetic polymers possessing high thermal stability have been obtained. These polymers can be used both in the form of soft and very hard materials.

Some advances have been achieved in the production of materials combining semiconducting properties with optical transparency and plasticity. Lunin et al. [6], for example, have reported the preparation of a plastic polymer with the electrical conductivity characteristics that are common to semiconducting materials. Substances promising for practical use can also be obtained by the action of ionizing radiations on polymers containing no conjugated bonds. A new semiconducting film has been developed [7] which will probably lead to an improvement in the design and technological production of articles of the electronic and electrotechnical industry. The film consists of an irradiated polyolefin with a resistance of 1600 Ω/cm². By bombardment with an electron beam, cross-linkages are formed in the polymer. The resulting material possesses the property of softening and of closely covering a surface of any configuration, retaining the shape imparted when the temperature is raised to 325°C. This property opens up new possibilities of creating conducting surfaces. Another property of the film is its capacity for self-sealing on being heated to 200°C.

In addition, the required shape can be obtained by stamping. The new material is already being studied as a screening coating for the protection of special high-voltage cables from coronary discharge (150 kV). This material may find wide use in the preparation of various electronic devices and electrical apparatus. The nominal working temperature of the material, at which it can be kept for a long time, is 105°C; however, in the absence of air it retains its properties at considerably higher temperatures.

The possibility of depositing semiconducting polymer coatings from solution is extremely attractive. Polyynes of the general formula $R_1-(-C\equiv C-)_n-R_2$ subjected to irradiation or to heating to a temperature below the melting point may be used as such polymers [8]. The initial polyyne is generally dissolved in acetone and the resulting solution is applied to a solid substrate, after which the solvent is evaporated. Semiconducting properties are imparted to the polyyne by heating or irradiation. Resistors, valves, and photoelements can be made from this substance.

The very name "organic semiconductors" led the first workers to attempt to use these materials in the fields of radioelectronics in which inorganic semiconductors have found such wide application. Several attempts were made to prepare elements possessing unsymmetrical static volt–ampere characteristics and, consequently, rectifying variable current. In the first [9] films of Saran (a copolymer of poly(vinylidene chloride) and poly(vinyl chloride)) was irradiated with far-ultraviolet radiation leading to a 1000-fold decrease in the dark resistance. At the boundary between the irradiated and the unirradiated surface of the film a p–n junction was formed which exhibited rectifying properties on irradiation with far-ultraviolet radiation.

Vannikov [10] showed that the modification of polyethylene gave materials with conductivity of the p-type. It was found possible by treating these materials with iodine under certain conditions to change the type of conductivity. In the case of tableted initial samples of p-type, iodination created an n-layer. However, in the samples obtained in this way the rectification factor did not exceed 25.

The possibility of obtaining a p–n junction in fibers of thermolyzed polyacrylonitrile has been reported by Kustanovich et al. [11, 12].

Rectifying junctions can be prepared by using two layers of low-molecular-weight organic semiconductors [13]. The components of the systems in which the junctions are formed are indigo, chloranil, phenazine, and a chloranil-phenylenediamine complex. Thin layers are deposited by vaporizing the substances in vacuum. Rectifying factors of the order of 20 with direct currents equal to several tenths of a microampere are found. It is possible that a p–n junction is obtained only in the phenazine-chloranil-phenylenediamine complex system and in all the other cases a $p-p^{+}$ junction is formed.

What has been said above permits the conclusion that the achievements in the field of creation of p–n junctions in organic semiconductors are extremely slight. However, this is partially compensated by the fact that at the present time it is possible to obtain polymeric films of practically any thickness with a conductivity from that of a dielectric to that of a semimetal (see, for example [14-16]). Because of this, the problem of the creation of p–n junctions has lost its urgency at the present time since active elements in the form of films can be created without the use of p–n junctions.

Investigations on the production of active elements in the form of films based on various principles, including the principle of space-charge-limited currents and the field effect, have acquired wide development. The basic requirements for the material of the semiconducting film will be as follows: fairly high mobility of the current carriers; low concentration of defects that could serve as electron-trapping centers; low concentration of surface states at the boundary with the dielectric; great width of the forbidden gap [17].

The possible methods for constructing active film elements have been mentioned in a study mainly of low-molecular-weight organic semiconductors. As early as 1960-1963, Kallmann and Pope [18, 19] showed that a solution of iodine in NaI + H_2O is an electrode that injects holes into a crystal of anthracene. If the other electrode is not an injecting electrode, the direct and reverse currents differ sharply in magnitude. It has been found [20] that a solid electrode of CuI deposited by vaporization also injects holes. If the second electrode is an aluminum one, sharply unsymmetrical volt–ampere characteristics are observed. For direct currents greater than 10^{-8} A the rectification factor reaches 10^3-10^5.

Rectification of alternating current for a pair of different electrodes has also been observed on the phthalocyanines of copper, nickel, and molybdenum and metal-free cyanine [21] and also for poly(ferrocene ketone) and poly(tetrachlorophenyl sulfide) [5]. A patent [22] describes a method for preparing diodes from organic semiconducting materials. The method of preparing the diodes from phthalocyanines and polyphthalocyanines is given in detail. In addition, nonlinear effects can be obtained by using saturated polymers such as polyethylene [23]. For this purpose, crystals of polyethylene with an area of $10 \times 20\ \mu^2$ and 100 Å thick are placed between a pair of electrodes — Cu/Pt, Mo, and W. At 500 mV the rectification factor reaches 25 at a direct current of ~25 mA. The fact deserves special mention that with a further increase in the voltage a region of negative resistance is observed on the volt–ampere curve. Organic diodes have been studied for the creation of a computer made completely with microelectronic circuits [24]. Large diode matrices acting as associative long-term memory devices can be prepared economically by means of a method permitting the formation of the diodes simultaneously for the whole matrix. This has been achieved by the vacuum vapor-deposition of copper phthalocyanine. The main advantage of copper phthalocyanine is the possibility of depositing it on flexible plates. The film of organic semiconductor deposited on the substrate was placed between cathode and gold anodes common for all the diodes arranged at the periphery of the film. As a rule, diodes on cards had an area of 3.5 mm^2 and were characterized by a rectification factor of $\sim 10^5$, a capacitance of 50–100 pF, and a direct current of 2 mA at a voltage of 1.5 V. However, the lack of reproducibility of these characteristics is reported [24].

Not only diode characteristics but also symmetrical volt–ampere characteristics with a nonlinear dependence on the current and the voltage are of interest for the creation of nonlinear semiconducting resistors. Nonlinear symmetrical volt–ampere characteristics can be found, in particular, in the case of currents limited by a space charge. Space-charge-limited currents have been studied in several organic semiconductors. An exponential dependence of the current on the voltage in vapor-deposited films of copper phthalocyanine has been observed [25]. If metals with different electron work functions are used as contacts, a rectifying effect is observed, the rectifying factor depending on the nature of the metal present in the phthalocyanine chelate complex and on the electrodes used [26].

The most important parameter characterizing a nonlinear resistor is the coefficient of nonlinearity β, which is defined as $\beta = (V/i)di/dV$. In space resistors it is usually 5 or less. At the present time it is possible to make low-voltage nonlinear film resistors from organic polymeric semiconductors with $\beta = 8$ and a working current of 2 mA at a voltage of 5–10 V. Such characteristics satisfy the requirements set for such resistors [27]; however, the use of polymeric materials is opposed by the poor stability of the specimens [28, 29].

The dependence of the conductivity of organic semiconducting materials on the temperature can be used to obtain thermoresistors — devices in which markedly nonlinear volt–ampere characteristics are observed because of the dispersion of power in the sample on the passage of an electric current. Thermoresistors can be used to measure and control temperatures, to stabilize voltages, etc. Data on the technology of the preparation of thermoresistors based on cyclic polynitriles and the characteristics of laboratory specimens have been given by Krivonosov et al [30].

Effects changing the electrical characteristics of polymeric semiconductors under the action of external mechanical forces may find actual use. Satisfactory results have been obtained in a study of the characteristics of the piezoresistance of certain highly conjugated polymers with semiconducting properties [31]. The possibility of the use of polymeric semiconductors as strain-sensitive sensors has been mentioned [32]. The widely used wire strain gauges have a coefficient of strain sensitivity $K = (\Delta R/R) l/\Delta l$ (where $\Delta R/R$ is the relative change in the resistance and $\Delta l/l$ is the relative deformation) between 1.9 and 2.1.

Recently, semiconducting strain gauges having enormously greater strain-sensitivity coefficients (of the order of 100) and high output signals not requiring complex apparatus for their measurement have begun to be widely used. But even in comparison with devices based on inorganic semiconductors, strain gauges made from polymeric semiconductors have a number of advantages, consisting in their lower sensitivity to temperature changes, stability to radiation, and simplicity of manufacture. The results of tests of strain gauges made from polycyanamide and poly(cyanic acid) are given in Figs. 43 and 44.

As can be seen from Fig. 43, the coefficient of strain sensitivity of an organic semiconducting sensor, defined as the angle of slope at any point of the curve, is 1.5–2 times this coefficient for inorganic semiconducting strain gauges, having a value of the order of 160–200. Moreover, as follows from Fig. 44, the temperature error of an organic strain gauge is considerably lower than that of an inorganic one. With an increase in the range of working temperatures (up to 800°C), the relative temperature error of a polymeric strain gauge becomes still less. A high coefficient of strain sensitivity, several orders of magnitude greater than the coefficient of strain sensitivity of metals, has also been observed in samples of thermolyzed polyacrylonitrile [33].

There is a report [34] on the possibility of using products of the reaction of tetracyanoethylene with salts of various metals to create thermoelectric devices. These compounds, which can be obtained in the form of discs, bars, strips, etc., are suitable for

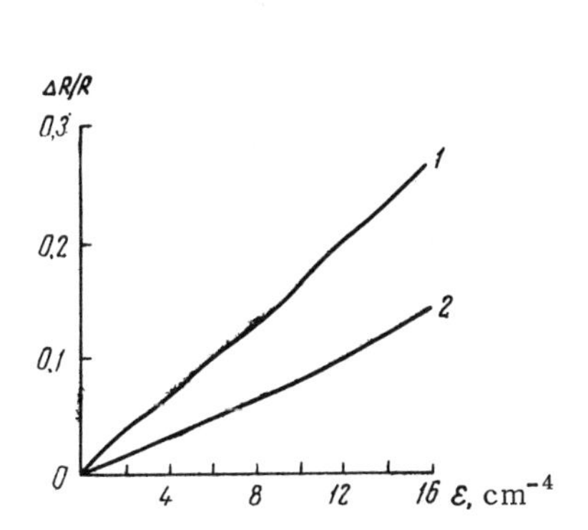

Fig. 43. Relative change in resistance ΔR/R as a function of the deformation ε. 1) Poly(cyanic) acid strain gauge; 2) inorganic semiconducting strain gauge.

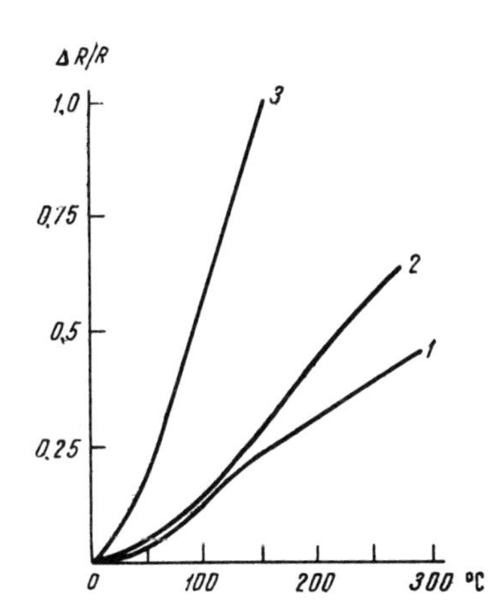

Fig. 44. Relative change in resistance ΔR/R as a function of the temperature. 1) Low-ohmic sample (45 Ω) of poly(cyanic acid); 2) high-ohmic sample (3660 Ω) of polycyanamide; 3) inorganic semiconducting sensor of the n-type.

the manufacture of thermoelectric energy generators and devices for thermoelectric cooling.

USE OF THE PHOTOELECTRIC AND OPTICAL PROPERTIES OF ORGANIC SEMICONDUCTORS

Let us consider the possibility of applications of organic semiconductors making use of their photoelectric and optical properties.

As early as 1954 [35] a suggestion was put forward that highly conjugated systems characterized by high absorption coefficients in the infrared region could be used as detectors of infrared radiation right down to the far IR region in which, as is well known, no reliable detectors have so far been found. The possibilities of using organic semiconductors have been considered by Drechsel and Görlich [36]. For practical use, two phenomena considered in Chapter II are important: the considerable photoelectric sensiti-

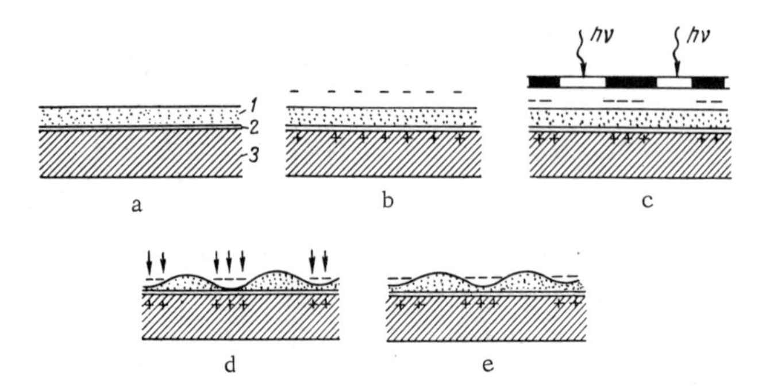

Fig. 45. Process of recording by means of photoconducting thermoplastic material. 1) Photosensitive thermoplastic layer; 2) conducting transparent layer; 3) substrate.

vity of polymeric charge-transfer complexes and the sensitization of the photoconductivity of polymeric semiconductors by organic dyes. Their use makes it possible to obtain photocurrents of considerable magnitude and to extend the spectral range of photoelectric sensitivity. Both these effects are used in the creation of various types of photoelectric devices especially in processes of electrophotography where considerable advances have been achieved at the present time [37].

Attempts were made long ago to use the effect of the interaction of electrostatic charges with insulating films to obtain images. One of the systems acting on this principle was first proposed in 1953 [38]. However, this system required the use of a vacuum, since the film was charged by means of an electric beam. This led to the deformation of the film by electrostatic forces even at room temperature. In another system [39] a thermoplastic film was used. In order to obtain an image it was necessary to heat the plastic leading to deformation of the film under the action of the electrostatic charges. The image could be frozen by rapid cooling. Organic semiconductors have found use in devices of this latter type [40].

We may mention the basic features of the method of electrophotography. Figure 45a shows the structure of an electrophotosensitive plate. The top layer consists of a photosensitive thermoplastic material, then there is a conducting transparent layer, for example gold, copper, or tin oxide. This conducting layer may

a

b

Fig. 46. Original image (a) and image obtained by the thermoplastic process (b).

also be reflecting, for example, vapor-deposited aluminum. Finally, the substrate may be made of glass or polyester resin. Figure 45 (b, c, d, e) shows the process of obtaining an image. First the plate is charged with the aid of a corona discharge; the surface may be charged either positively or negatively. Charges of the opposite sign arise because of the image force in the conducting layer adjacent to the electrode. On illumination the charges disappear because of recombination at those sites where light falls, since the film is photoconducting; on heating, the film becomes plastic and the thermal development of the image takes place. In the final state the shaped profile of the image can be seen.

The quality of the reproduction of an image by the method considered can be evaluated from Fig. 46 [41], in which the resolving capacity of the material was 20 lines per mm. It is possible to use very diverse organic semiconductors as the photoconductors [40, 41]. Very many of them are transparent in the visible region and absorb in the ultraviolet region. In order to obtain a thermoplastic film the semiconductors are dissolved in polystyrene with a mean molecular weight of 20,000–30,000. The spectral characteristics of the conducting compounds can be improved by various additives. For example, the leuco base of malachite green has absorption peaks in the ultraviolet region (2080 and 2640 Å) but in practice the cationic form of the dye with absorption in the longer-wave part of the spectrum (3150, 4250, and 6150 Å) is always present in the leuco base. This cationic form sensitizes the layer of photoconductor in the visible region. In addition, such dyes as auramine, acridine orange, rhodamine B, etc., may be added specially.

A principle very similar to the principle of electrophotography is based on a television transmitting tube with a photoconducting target ("Vidicon"). The photoconducting layer must have a high quantum yield and be a dielectric in the dark.

Compounds with conjugated bonds have found wide use in copying devices. These include the diazotype process — the preparation of copies by the action of light without the use of silver-containing light-sensitive materials [42]. So-called photoresistors are also used in printing and radioelectronics. By means of photoresistors it is possible to obtain relief images or schemes in micromodular printing devices.

The reversible change in the absorption spectrum under the action of light apparently creates the possibility of using such compounds as working elements in optical computers [42]. Spiropyrans, for example, acquire a coloration under the action of ultraviolet radiation which can be retained for a long time. The appearance of a coloration is evidently due to an increase in the degree of conjugation, since under the action of light the molecule becomes coplanar as a consequence of the cleavage of a carbon–oxygen bond. The "readout of information" can be performed at those wavelengths that cause no changes in the dye. Light of another wavelength can decolorize the compound and thus erase the information recorded. Compounds exist which can be restored to their initial state not only by visible light but also by near-infrared radiation and also by a constant electric field [42].

The photochromic properties of organic substances can be used to make dosimeters of ultraviolet radiation and screens protecting the eyes and optical devices from the action of powerful sources of light (for example, in nuclear explosions).

Attempts have probably been made to use photochromic materials for tracking the movement of war materials [2], as masking dyes in changing illumination, and in artificial satellites with the object of stabilizing the temperature conditions [42].

In addition, attempts have been made to use photochromic systems in devices for converting light energy into electrical energy. In an irreversible photochemical reaction, up to 25% of the luminous energy can be transformed. However, the reduction of a leuco form by electrochemical oxidation decreases the efficiency of such systems to 1-2.5%. There is no doubt, however, that further progress in this field will depend on advances in the electrochemical investigation of such systems. Work in this direction is already being carried out [43], which must lead to a better understanding of the processes taking place in converters of light energy into electrical energy.

ORGANIC SEMICONDUCTORS AS ACTIVE MEDIA FOR LASERS

Organic semiconductors will probably find an important application as laser materials. In this case, by organic semiconductors we have in mind those organic compounds the molecules of which contain a system of conjugated double bonds. The intrinsic semiconducting properties of these compounds have not been studied, as a rule. At the present time two directions in studies of this field may be isolated.

The first of them is connected with the study of the induced radiation in chelate compounds of the rare-earth metals. In these compounds the induced radiation is caused by a system of levels of the metal ion. The organic moiety plays an auxiliary role. The second direction is connected with the possibility of obtaining coherent induced radiation actually in organic substances with conjugated double bonds. In this case, the induced radiation is due to electronic transitions in the organic molecule.

Chelates are molecules in which the central inorganic ion is bound by covalent bonds to one or more organic molecules. In these substances the desirable properties of the rare-earth ions and of systems of organic molecules are combined, namely: the fluorescence spectra of the rare-earth ions with their discrete narrow lines and the wide absorption band of the organic molecules in the ultraviolet region covering a considerable part of the visible region. This association of properties greatly increases the effi-

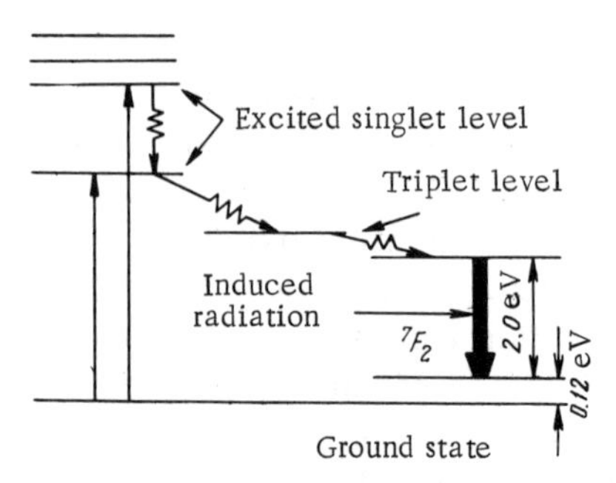

Fig. 47. Scheme of the electronic transitions in europium chelates. ∧∧∧→ radiationless transition; → radiation-emitting transition.

ciency of the use of the wide-band light of the pumping from the usual flash tube as compared with what can be obtained with ions of the rare-earth elements alone with their narrow absorption lines.

Figure 47 shows a scheme of the action of a chelate laser using europium tribenzoylacetonate, which is as follows [44]. The molecule of europium tribenzoylacetonate consists of an europium ion and three benzoylacetonate radicals attached to it. By absorbing the energy pumped in, the molecule passes from the ground state into the excited singlet state. From the first excited singlet state the molecule passes without radiation into the lower triplet biradical state, which is characterized by the fact that the spins of the two electrons are not coupled. Since the triplet level is above the resonance level of the central ion, the energy is transferred to this level and the rare-metal ion emits fluorescence lines. The very intense fluorescence line at 613 nm corresponds to the transition to the final $^7F_2$ level which lies 0.12 eV above the ground state. This energy is so large that the equilibrium population of the final level is ~1% of the population of the ground level at room temperature and decreases greatly when the temperature is lowered. Consequently, the chelates function in the manner of materials used for a four-level laser.

To the question of whether conjugated double bonds are necessary for such a system, we must answer in the affirmative since, in the first place, the chelate compound can be formed only when conjugated double bonds are present and, in the second place, the organic moiety of the chelate molecule must have a sufficiently broad optical absorption band lying in the ultraviolet region for it effectively to absorb the light of the pump.

Thus, for example, for the mercury lamps frequently used the substance must possess absorption capacity in the 365 nm region. Ordinary saturated organic compounds begin to absorb intensively only in the vacuum ultraviolet and, of course, are unsuitable for the production of laser media since excitation with vacuum ultraviolet is associated with great difficulties. In molecules containing conjugated double bonds, thanks to the interaction of the π-electrons the excitation levels are greatly lowered and these substances absorb not only ultraviolet but also visible light. Let us give some typical characteristics of lasers based on europium benzoylacetonate.

When europium benzoylacetonate was introduced into a polymeric matrix [45], the radiation line at 77°K had a central wavelength of 613 nm, the width of this line was 15 Å, the lifetime was 0.5 μsec, and the quantum efficiency reached 80%. With these parameters, the threshold energy of the pumping of the laser is 0.01 J and a low-power source can be used for pumping.

The generation of stimulated radiation has been achieved [46] in frozen vitreous solutions of europium benzoylacetonate (with a concentration of 5.2×10^{18} molecules/cm^3) in a mixture (3 : 1) of ethanol and methanol. The luminescence of the solutions at 100–150°K consists mainly of the lines 6130 and 6150 Å (widths 8 and 20 Å). The duration of the luminescence is 5×10^{-4} sec. The organic moiety of the complex absorbs in the 390–420 nm region. Generation took place at an input of 190 J. With a pumping energy exceeding the threshhold value by 30%, the half-width of the generated line (613 nm) was 0.3 Å.

The organic moiety of a chelate molecule may be of different types. Investigations are being carried out on the productions of lasers based on chelates the organic moiety of which consists of long molecules of trifluorophenylacetone. Optical pumping is effected by the use of the absorption band of the organic moiety of the molecule with a wavelength of 340 nm. The laser transmission arises close to 613 nm. The time of fluorescence at 77°K is 1.5 μsec.

Even a cursory consideration of the results of investigations on stimulated radiation in chelates gives grounds for concluding that these investigations are now being carried out at the level of design studies. The production of a laser transition in liquids and glasses with the addition of chelates as active materials brings to actual fruition the idea of the use of a matrix of various polymeric materials with valuable design characteristics. In actual fact, there are reports on the making of a laser [45] which consists of a spiral of poly(methyl methacrylate) fiber containing europium benzoylacetonate. The laser works on cooling to 77°K. Pumping is effected by a pulsed xenon lamp with radiation at 340 nm. The laser gives a pulse of crimson light at 613 nm with a duration of ~250 μsec.

The use of polymers opens up the possibility of imparting very diverse forms to the active media of lasers. In particular, the active medium of a laser based on poly(methyl methacrylate) containing europium chelates can be made in the form of a thin sheet or threads with the thickness of a human hair [47].

Lasers based on chelate compounds possess a whole series of properties which, in the majority of cases, have advantages in comparison with the properties of ordinary lasers [48].

These advantages are, first, in the good solubility of the chelates in many organic solvents and various plastics, as a result of which the latter can be given any form appropriate to special purposes; second, in the possibility of obtaining lasers based on liquids, which eliminates the problems connected with the growing of single crystals and their treatment, since with liquids it is possible to obtain considerable concentrations of active centers in the working volume and large samples of the required dimensions; third, in the fact that the solution of the cooling problem is provided by the possibility of circulating the working liquid. The vessel containing the liquid may have inlet and outlet apertures enabling the liquid to be circulated between the laser and the cooling device. The rapid removal of heat is necessary for high-powered lasers.

Now let us turn to the question of the creation of lasers based on conjugated systems of double bonds in organic compounds as such. The possibility of using organic compounds for the generation of coherent radiation was shown by Brock [49].

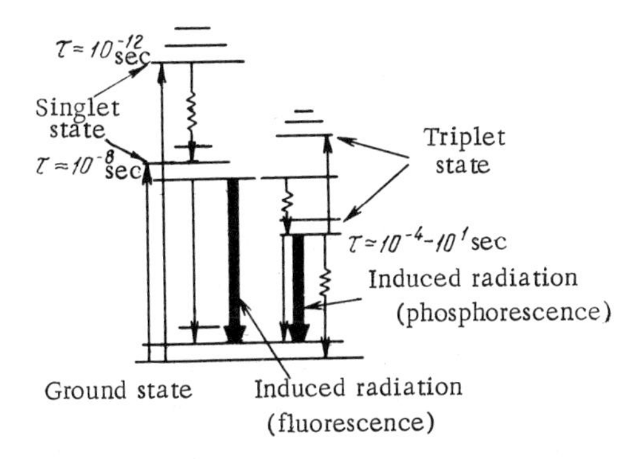

Fig. 48. Scheme of the electronic transitions in aromatic hydrocarbons. ⇝ Radiationless transition; → radiation-emitting transition.

The processes taking place in an organic molecule and shown in Fig. 48 are analogous to the processes taking place in the three-level system of a ruby laser and consist in the transition of electrons from the ground singlet level to the first and higher excited singlet levels under the action of the exciting light. Then the radiationless transition of the electrons from the excited singlet level to the lower metastable triplet level which is particularly characteristic for organic substances takes place.

The observed induced radiation has been ascribed [50, 51] to the transition of an excited electron from the triplet level to the ground level, i.e., the laser effect was connected with the phosphorescence of the organic substances. As active materials solutions of acetophenone, benzophenone, pyrazine, and α-bromonaphthalene in a mixture of methylcyclohexane and isopentane (1 : 1 or 1: 2) with a molar concentration of $\sim 10^{-3}$ mole/liter were used.

Laser radiation has been obtained [52] in indole distributed in a vitreous matrix consisting of a mixture of diethyl ether, isopentane, and ethanol at 77°K. Oscillograms with spattering in the blue part of the spectrum were observed. To investigate promising media for organic lasers, the phosphorescence spectra of a number of aromatic hydrocarbons introduced into various plastics, boric acid, and glucose have been studied [44, 53]. The results of the investigation show the possibility of making on the basis of such media lasers that work at room temperature and require small pumping energies. Finally, yet another possibility of obtaining coherent radiation from organic substances has been established; namely, by stimulated Raman scattering [54]. The beam of a ruby laser may be used as the source of luminous pumping. At the present time lasers are being developed on the basis of organic materials working over a wide range of wavelengths [55]. Generation has been achieved in the near- and far- IR region with benzene, nitrobenzene, deuterated benzene, cyclohexane, pyridine, and other substances. Generation arose when these liquids were excited with the light of a very-high-powered ruby laser. The stimulated emission of the Raman light took place with an intensity of up to 30% of the intensity of the exciting radiation. Such systems can be considered as converters of the frequency of the coherent radiation of a ruby laser into a whole set of other frequencies.

It must be observed that the study of organic compounds to determine their suitability as active media for laser devices presents fundamental difficulties due both to the impossibility of accurately characterizing the purity of the organic substance and to the necessity of including new methods differing from those usually used for the study of coherent induced radiation in the case of inorganic crystals. Consequently, it is not very surprising that the results of the work of Morantz et al. [50] were not reproduced in similar investigations [56, 57]. Nevertheless, interest in organic substances from the point of view of their use as laser materials is not slackening since they possess such valuable characteristics as high power, wide selection of generated wavelengths, possibility of chemical tuning, simplicity, and high efficiency.

Let us consider various processes of spontaneous radiation taking place in aromatic organic compounds [58]. The molecular energy levels of aromatic molecules are fairly diffuse, in contrast to the sharp energy levels characteristic for atoms. Transitions between such levels give relatively wide bands in absorption and emission the width of which can vary from values of less than 100 Å up to about 1000 Å. The excited states with the different lifetimes by which they are characterized in aromatic hydrocarbons (where only $\pi - \pi^*$ transitions exist) can be discussed with the aid of the diagram shown in Fig. 48. It can be seen from the figure that in principle generation is possible both in phosphorescing and in fluorescing organic aromatic substances.

The main criterion for the selection of suitable organic compounds for lasers may be taken as the number of molecules required to start generation in the optical resonator:

$$\frac{\Delta N}{V}=\frac{4\pi^2 c\Delta\bar{\nu}\alpha\tau_{spont}}{(\pi\ln 2)^{1/2}L\mu\cdot\lambda^2_{em}}, \qquad (1)$$

where $\Delta N/V$ is the difference in the "populations" of the corresponding generating levels in unit volume, τ_{spont} is the intrinsic radiative life, α represents the total losses in one passage in the generating medium in the resonator, L is the length of the resonator, μ is the refractive index, $\Delta\bar{\nu}$ is the width of the fluorescence band at half height expressed in wave numbers, λ_{em} is the wavelength of the radiation emitted, and c is the velocity of light.

A promising four-level fluorescing system for the creation of a laser is the perylene molecule, which is photochemically stable under the conditions of intensive optical irradiation. In a solution of perylene in a liquid or solid solvent the energy of the exciting light is used more effectively because of the intermolecular transfer of the energy absorbed by the molecules of the solvent to the radiating centers. The following figures have been obtained for a solution of perylene in benzene [59]: λ_{em} = 4710 Å; $\Delta\bar{\nu}$ = 2.12×10^4 cm^{-1}; λ_{exc} = 4100 Å; $\tau_{spont} = 6.9 \times 10^{-9}$ sec at a quantum yield of 0.96; μ = 1.5; L = 6 cm; and α = 0.18 (on the assumption that scattering is the only appreciable form of loss). From formula (1) we find that $\Delta N/V = 4.3 \times 10^{13}$ cm^{-3}. However, this value is too low, since it is assumed that all the radiation is concentrated in the band at ~4710 Å, while in actual fact only one-third of the total emission is found in this region. Furthermore, it was assumed that the final state was thermally depleted, although the actual distance between the final ground states is 1500 cm^{-1}. In view of this, taking the initial concentration of perylene as 6×10^{16} cm^{-3} and calculating the concentration of molecules in the final generating state we obtain $\Delta N/V = 1.7 \times 10^{14}$ cm^{-3}.

The minimum critical pumping power for generation in this system is given by the formula

$$P_{crit}=\frac{\Delta Nhc}{V\lambda_{exc}\tau_{spont}}=1.3\cdot 10^4\ \mathrm{W/cm^3} \qquad (2)$$

It is interesting to note that for a ruby laser this magnitude is less by half an order of magnitude, amounting to 6×10^2 W/cm^3. It is possible to evaluate the required total electrical power of the source of light for exciting generation. Assuming that for a source with a broad spectrum (for example, a xenon pulse lamp) the probability of conversion is 1% and that the probability of the selection and absorption of radiation is 50%, we obtain an electrical power of 2.6×10^6 W/cm^3. For a volume of 3 cm^3 and with a light pulse lasting 0.1 μsec an electrical energy of excitation of ~8 J is required. At the present time such sources are fully available.

Analogous considerations for a three-level system leads to less encouraging results. Since the final state coincides with the ground state, at the same initial concentration of molecules of perylene the quantity $\Delta N/V \approx 6 \times 10^{16}$ cm^{-3} and the critical pumping power, and also the electrical power of the source of light, rises by more than two orders of magnitude. Such powers are apparently difficult to achieve with modern optical sources of excitation. However, they can be obtained in the electron pulse of an electron accelerator.

The situation is worse with fluorescing aromatic substances. As has been shown, the difference in the filling of the corresponding generating levels is proportional to the intrinsic radiative life and it is several orders of magnitude greater in phosphorescing substances than in fluorescing substances. This leads to the situation that, for example, for benzophenone — a four-level phosphorescing compound — with the achievement of triplet–triplet absorption the threshold of the inversion of the "population" is estimated at 3×10^{20} cm^{-3}. It is quite clear that at the present time such systems are unpromising for use as laser materials.

A simple consideration of the suitability of various aromatic organic substances for obtaining induced irradiation makes it possible to conclude that the generation of coherent radiation can be obtained with aromatic molecules consisting of a four-level fluorescing system.

It can be seen from what has been said above that at the present time there are no investigations on the possibility of making lasers based on polymeric semiconductors. However, it can be said that the system of levels in polymeric semiconductors is far more diverse than in low-molecular-weight compounds with conjugated double bonds and, in particular, the problem of creating a laser emitting IR light is apparently easier to solve just with polymeric semiconductors.

THE SUPERCONDUCTIVITY AND WAVEGUIDE PROPERTIES OF MACROMOLECULES WITH CONJUGATED BONDS

Among the numerous problems arising in the study of the possibility of using organic semiconductors we may mention only two. The first of them relates to the possibility of synthesizing organic superconductors.

London long ago suggested the possibility of superconducting states in some large organic molecules of the protein type. Little [60, 61] has suggested that in addition to the known mechanism for the formation of the superconducting state in metals another mechanism is possible in which the temperature of the transition to the superconducting state may be considerably higher than in the cases so far known. To achieve the superconducting state it is necessary that the electrons that experience the force of Coulomb repulsion at ordinary temperatures be attracted to one another. In a superconductor, the attraction of the free electrons causes their joining in pairs. The attraction between the two similarly charged particles may arise as a consequence of the interaction of the electrons with the lattice. Then the gain in energy from the pairing of the two electrons exceeds the energy of the mutual repulsion of the particles. This is easy to visualize by considering two spheres present on an elastic film [62]. If the spheres are remote from one another, each of them creates its own depression in the film. When they approach to a certain distance, the two spheres will be present in a common depression. In a similar manner, each electron creates round itself a region where the lattice is distorted. When an electron moves, the "fur coat" that the electron pulls after it lags behind because the ions of the lattice move more slowly than electrons. Consequently, a second electron interacting with the first will move at some distance from the first. To understand how the superconducting state may be destroyed one must bear in mind the fact that in addition to these two electrons under consideration there are many others chaotically moving in all directions. These latter also cause a distortion of the lattice and if such "foreign" electrons pass close to the pair under consideration it may come about that the pair will separate. A necessary condition for this not to happen is the equality of the momenta of the centers of mass of all the electrons. This highly ordered state is possible only at low temperatures when the thermal motion is insufficient to destroy the pairs.

Other mechanisms can be imagined causing attraction between electrons and leading to the appearance of a superconducting state. One of such mechanisms was proposed by Little [60, 61]. Let there be a linear chain of conjugated bonds with symmetrically arranged lateral radicals which are capable of being polarized — for example, of the type of diethylcyanine iodide. When the electron moves along the main chain it polarizes the side-chains. However, the wave of polarization lags behind the movement of the electron thanks to which an electron moving after the first experiences attraction. Numerical estimates show the temperature of the superconducting transition is in the region of 2000°K.

Other effects connected with this type of interaction may also be expected, for example, the appearance of a negative dielectric constant [63]. The negative dielectric constant may serve as indication of the appearance of the superconducting state.

The work of Little has served as an impulse to the search for other possible mechanisms of the formation of a superconducting state. A scheme has been proposed [64] according to which surface superconductivity may arise as a consequence of the attraction between electrons in surface states. Such superconductivity does not require that the crystal itself be a conductor. To increase the attraction between electrons it is possible to use Little's mechanism by introducing strongly polarizing impurities into the crystal.

Up to now, there have been no experimental studies of polymers with the object of achieving the superconducting state. However, it is clear that in the study of polymeric materials measurements must be carried out with alternating current in order to exclude the influence of intermolecular barriers.

The second problem is connected with the waveguide properties of conjugated systems. It is not excluded that large polymeric molecules may be schematic elements of film electronics. It has been mentioned previously that it is possible to obtain stimulated emission in threads of chelate polymers. At the same time, polymeric chains may be waveguides for the propagation of light waves. By combining light waveguides, active chains with stimulated emission of radiation, and other photoactive elements it may be possible in the future to create basically new electronic devices [65]. The long molecules of polymers may also act as conductors for the propagation of electronic waves from one element of a scheme to another. The passage of electrons from one molecule to another may be controlled by an external field, by heating, and by the action of light. From this point of view the results of a study of the autoemission characteristics of a tungsten point coated with adsorbed layers of molecules with conjugated bonds are extremely interesting [66]. Further investigations in this direction may lead to the creation of molecular electronics based on the use of polymers with conjugated bonds.

REFERENCES

1. A. V. Topchiev (ed.), Organic Semiconductors [in Russian], Izd. Akad. Nauk SSSR (1963).
2. Y. Okamoto and W. Brenner, Organic Semiconductors, Reinhold, New York (1964).
3. V. V. Pen'kovskii, Usp. Khim., 33:1232 (1964).
4. M. Tomaino, Electronics, 36:36 (1963).
5. H. Pohl, in: Modern Aspects of the Vitreous State, J. D. Mackenzie (ed.), Butterworth, London, Vol. 2 (1962), p. 72; H. Pohl, Khim. i Tekhnol. Polimerov, 9:3 (1963).
6. A. F. Lunin, Ya. M. Paushkin, V. R. Mkrtchan, and N. A. Surovtsev, in: Semiconducting Polymers with Conjugated Bonds [in Russian], TsNIITÉneftekhim., Moscow (1966), p. 108.
7. Proc. Inst. Electr. Electronics Engrs., 53(1):6A (1965).
8. French Patent 1,274,975 (1962).
9. G. Oster, G. Oster, and M. Kryszewsky, Nature (London), 191:164 (1961).
10. A. V. Vannikov, Dokl. Akad. Nauk SSSR, 152:905 (1963).
11. I. M. Kustanovich, I. I. Patalakh, and L. S. Polak, Kinetika i Kataliz, 6:167 (1963).
12. I. M. Kustanovich, I. I. Patalakh, and L. S. Polak, Vysokomolek. Soedin., 6:197 (1964).
13. J. E. Meinhard, J. Appl. Phys., 35:3059 (1964).
14. A. A. Berlin, L. I. Boguslavskii, R. Kh. Burshtein, N. G. Matveeva, A. A. Sherle, and N. A. Shurmovskaya, Dokl. Akad. Nauk SSSR, 136:1127 (1961).

15. B. É. Davydov, Doctoral Thesis [in Russian], Inst. Neftekhimich. Sinteza Akad. Nauk SSSR, Moscow (1965).
16. A. V. Vannikov and N. A. Bakh, Élektrokhimiya, 1:617 (1965).
17. Questions of Film Electronics [in Russian], "Sovetskoe Radio," Moscow (1966), pp. 145, 207.
18. H. Kallmann and H. Pope, J. Chem. Phys., 38:2648 (1963).
19. H. Kallmann and H. Pope, Nature (London), 185:753 (1960).
20. I. A. Eligulashvili, G. A. Nakashidze, L. D. Rozenshtein, and A. A. Khadiashvili, Élektrokhimiya, 2:107 (1966).
21. F. Haak and J. Notla, J. Chem. Phys., 38:2648 (1963).
22. German Democratic Republic Patent 3337 (1964).
23. A. Van Roggen, Phys. Rev. Letters, 9:368 (1962).
24. M. Wolff, Electronics, 36:35 (1963).
25. G. Delacote, I. P. Fillard, and F. J. Marco, Sol. State Comm., 2:373 (1964).
26. Z. A. Rotenberg, S. D. Levina, and L. A. Korob, Élektrokhimiya, 2:1224 (1966).
27. M. M. Nekrasov, Microminiaturization and Microelectronics Based on Nonlinear Resistances [in Russian], "Sovetskoe Radio," Moscow (1965).
28. L. I. Boguslavskii and L. S. Stil'bans, Dokl. Akad. Nauk SSSR, 147:1114 (1962).
29. A. V. Vannikov, V. I. Zolotarevskii, and D. I. Naryadchikov, Élektrokhimiya, 3:1379 (1967).
30. A. K. Krivonosov, A. F. Lunin, Ya. M. Paushkin, and V. I. Ruslanov, in: Semiconducting Polymers with Conjugated Bonds [in Russian], TsNIITÉneftekhim., Moscow (1966), p. 148.
31. A. W. Herny and G. Cappas, U.S. Dept. Com. Office Techn. Serv., AD 403:361, 40 pp (1963); Chem. Abstr., 60:6305 (1964).
32. Ya. M. Paushkin, I. I. Krivonosov, A. F. Lunin, and V. R. Mkrtchan, in: Semiconducting Polymers with Conjugated Bonds [in Russian], TsNIITÉneftekhim., Moscow (1966), p. 151.
33. A. A. Averkin, A. V. Airapetyants, Yu. V. Ilisavskii, É. L. Lutsenko, and V. S. Serebrennikov, Dokl. Akad. Nauk SSSR, 152:1140 (1963).
34. J. E. Katon, US Patent 3,267,115 (1966).
35. B. McMichael, E. Kmetko, and S. Mrozowsky, J. Opt. Soc. Am., 44:26 (1954).
36. L. Drechsel and P. Görlich, Infrared Phys., 3:229 (1963).
37. I. B. Sidaravichyus, Voprosy Radioélektroniki, 1965(12):105.
38. E. I. Sponable, J. Soc. Motion Pict. Technicians, 60:337 (1953).
39. W. E. Glenn, J. Appl. Phys., 30:1870 (1959).
40. H. G. Greig, RCA Review, 23:413 (1962).
41. N. E. Wolff, RCA Review, 25:200 (1964).
42. Yu. N. Gerulaitis and A. I. Korolev, Zh. Vses. Khim. Obshchestva im D. I. Mendeleeva, 9:78 (1966).
43. K. Nakada and M. Sato, J. Electrochem. Soc. Japan, 34:28 (1966).
44. Electronics, 36(15):6; (17):11; (46):38 (1963).
45. Electron. Design, 11(6):18 (1963).
46. A. Lempicki and H. Samelson, Phys. Letters, 4(2):133 (1963).
47. Steel, 152(12):76 (1963).
48. B. Lengyel, Lasers, Wiley, New York (1962).

49. E. G. Brock, J. Chem. Phys., 35:759 (1961).
50. D. J. Morantz, B. C. White, A. J. Wright, Phys. Rev. Letter, 8:23 (1962); Proc. Chem. Soc., 1962:26; J. Chem. Phys., 37:2041 (1962).
51. J. Macauley, Electron. News, 7(34):127 (1962).
52. Electronics, 35(35):9 (1962).
53. G. Oster, N. Geacintov, and A. U. Khan, Nature (London), 196:1089 (1962).
54. M. L. Kats, M. A. Kovner, and N. K. Sidorov, Lasers [in Russian], Izd. Saratovsk. Gos. Univ., Saratov (1964), p. 270.
55. Electronics, 34(4):7 (1962).
56. A. Lempicki and H. Samelson, Appl. Phys. Letters, 2:159 (1963).
57. F. Wilkinson and E. Smith, Nature (London), 199:631 (1963).
58. D. L. Stockman, W. R. Mallory, and K. G. Tittel, Proc. Inst. Electr. Electronics Engrs., 52(3):341 (1964).
59. J. Bowen and R. J. Leginston, J. Am. Chem. Soc., 76:6300 (1954).
60. W. A. Little, Phys. Rev., 134:A1416 (1964).
61. W. A. Little, Scientific Am., 212(2):21 (1965).
62. W. Little, Usp. Fiz. Nauk, 86:317 (1965).
63. L. V. Keldysh, Usp, Fiz. Nauk, 86:327 (1965).
64. V. A. Ginzberg and D. A. Kirzhenets, Zh. Éksperim. Teoret. Fiz., 46:387 (1964).
65. N. G. Nakhodkin, in: "The Physics of Metallic Films" [in Russian], "Naukova Dumka," Kiev (1965), p. 4.
66. A. P. Komar and A. A. Komar, Zh. Tekhn. Fiz., 31:231 (1961).

SUBSTANCE INDEX

SUBJECT INDEX